D0991412

COLONIZING OTHER WORLDS

A Field Manual

Other books by the author

INTERSTELLAR TRAVEL
ALONE IN THE UNIVERSE?
JOURNEY TO ALPHA CENTAURI
(published in paperback as
HOW WE WILL REACH THE STARS)
WHISPERS FROM SPACE
SPACE WEAPONS/SPACE WAR
WHERE WILL WE GO WHEN THE SUN DIES?

COLONIZING OTHER WORLDS

A Field Manual

JOHN W. MACVEY

STEIN AND DAY/*Publishers*/New York

First published in 1984

Designed by Louis A. Ditizio
Printed in the United States of America
STEIN AND DAY/*Publishers*
Scarborough House
Briarcliff Manor, N.Y. 10510

Library of Congress Cataloging in Publication Data

Macvey, John W.
Colonizing other worlds.

Bibliography: p.
Includes index.
1. Space colonies. I. Title.
TL795.7.M33 1984 919.9'04 83-40006
ISBN 0-8128-2943-3

To America's first woman
in space,
astronaut Sally Ride

Acknowledgments

Once again it is my very pleasant duty to acknowledge with sincere thanks all those who assisted in the preparation of this book. These include Mrs. Helen Duncan, who has again been able to transform my feeble attempts at artistry into excellent diagrams, and my younger daughter, Karen, who has had, over several months, to cope with her father's barely legible scrawl and transform this into a respectable typescript.

Last and certainly by no means least, I would like to express my sincere thanks to Benton Arnovitz, the editorial staff, and all other members of Stein and Day who have at all times given their utmost encouragement, assistance, and advice.

John W. Macvey

Contents

Witness this new-made World, another Heaven
From Heaven-gate not far, founded in view
On the clear hyaline, the glassy sea;
of amplitude almost immense with stars
Numerous, and every star perhaps a world
of destined habitation . . .

Milton, *Paradise Lost*

Preface

In my last book, *Where Will We Go When the Sun Dies?* I dealt at length with the reasons why it will one day be imperative for terrestrial civilization to quit the Solar System in favor of an Earth-like planet orbiting a star similar in type and age to today's Sun. What motivated that book was the prospect of some predictable catastrophe about to overtake the Sun sometime in the next five billion years. But if as many as possible are saved from that coming holocaust and taken to a suitable planet orbiting a safe star, only two-thirds of the problem will be solved. There will remain the gargantuan task of reestablishing civilization in an alien environment. Clearly the problems will be tremendous, but if all or at least a major portion of the human race supports the project, the refugee fleet of interstellar transports (ISTs) would contain vast quantities of the supplies and equipment essential to enable a sound start to be made. Whether a project on this scale ever *can* be fully achieved remains moot. It may be possible only if the world's many jealous governments and peoples pool all their resources. Since agreement among nations is not a feature that we particularly associate with our planet, it is more likely that mankind instead will see individual nations building their own ships and striving to save as many of their own people as possible—and probably fighting with each other at the same time over the necessary materials and provisions. It would be much more pleasant to visualize the nations of Earth pulling together in the face of a common catastrophe, but it is difficult to be sanguine in the face of our history.

I began to think in terms of a scaled-down version of such a project, a much smaller one, just a few ISTs and a limited number of passengers. Moreover, I put the idea of a dying Sun behind me and easily discerned other reasons why such an expedition might be mounted—reasons likely to come into play much, much sooner than the death of the Sun. Among those reasons are a need to escape from political or sociological pressure on Earth or to flee a planet polluted, overcrowded, and suffering increasing shortages of food, fuel, minerals, and other raw materials. These two evils are far from hypothetical. The first has been an obvious danger since the dawn of the twentieth century and perhaps before; the second is presently the more sinister because, though its effects are already apparent, its full implications have not yet begun to sink in. By the time some form of interstellar travel is perfected it should be obvious to everyone! A third reason is equally valid and should be equally obvious—the very real risk of nuclear Armageddon. We can only pray that our lifeline to the stars will exist before then.

The basic scenario of this book will focus on the idea of a colonizing expedition to the stars, a venture founded for one or another of the reasons stated above. Because of the awesome parameters of time and distance involved, we will also adhere strictly to the concept of a one-way journey. To complete the scenario it will be necessary to include one further important factor. In my book *Interstellar Travel: Past, Present, and Future* I endeavored to envisage and predict methods of star travel of an essentially unique kind—those that would not involve the pathos and poignancies of the more conventionally envisaged techniques. These invoked travel via possible shortcuts through curved space or through black holes, to mention but two. However, such techniques are still, at best, only theoretical, and it seems very much more likely that the first interstellar expeditions from Earth will be compelled to fall back on much less sophisticated techniques such as generation travel (the well-known "space ark" concept) or the use of cryogenics to suspend the aging process as the ISTs traverse the light-years of space. In a still earlier book, *Journey to Alpha Centauri,* written in 1964, I employed the "space ark" technique whereby seven generations in a pair of vast ISTs traveled to a world orbiting one of the twin suns of the nearest star to our Sun, Alpha Centauri, in order that the eighth could take over that world and set up a new and better terrestrial civilization.

In the present scenario, however, the men and women quitting Earth are the same men and women who will reach the new world. This, almost of necessity, invokes cryogenic techniques by which human beings are put into a form of protracted coma and then more or less deep-frozen; their life processes just tick slowly on until an automatic warning system causes them to awaken when their fully automated IST is within reach of the destination star system. Some progress is indeed being made in the science of cryogenics, but I am not for a moment suggesting that it has yet reached the stage of rendering interstar journeys possible.

At the moment this seems the most promising prospect. Whether it will prove to be practical, only time will tell. I doubt very much that by then I will be around to find out.

A colonizing expedition on these lines will be very much on its own. If we had envisaged some rapid-transit, extradimensional technique, we could equally well have envisaged a shuttle system of space vehicles maintaining a constant flow of traffic between Earth and the "new" planet, bearing all the supplies and reinforcements required. But this would hardly have been in accord with a *Mayflower* type of voyage. Just as the early pioneers who landed on Plymouth Rock were on their own, so would be the colonists we are considering in these pages. As a consequence there are a multitude of factors that the organizers of such an expedition would have to recognize, a host of problems to which they would have to find the right answer, for there could be no second chance, no possibility whatsoever of sending one of the ISTs back to Earth to pick up some essential item or commodity that had been overlooked. Such a colony, light-years deep in space, could eventually perish. Equally it could lay the foundations of a new, more just, and greater terrestrial civilization. Much would depend on the original degree of planning, the wisdom or otherwise of the decisions made. At the same time, it is essential to remember that other parameter that can never be guaranteed. It comes under different names—luck, chance, fortune, fate, the will of Providence. However, we will assume in the pages to follow that fate favors the pioneers by virtue of sound planning and wise decisions. We will try to envisage the many dangers and difficulties that must prove inevitable. It is the study of these with which this book is primarily concerned.

COLONIZING OTHER WORLDS

A Field Manual

1

Why Leave Earth?

BEFORE EMBARKING ON the planning and execution of a one-way voyage to the world of another star, we really ought to decide why that trip will be either necessary or desirable. Travel to another star invokes parameters of time and distance that divorce it entirely from journeys between Earth and the other planets of our Solar System and, as we know, even these are still extremely daunting and will remain so for many decades. It would be easy, of course, to fall back on an old cliché that still retains a fair measure of relevance. Why climb Mount Everest? Because it is there! Why go to the Moon? Because it is there! Why go to the stars? Because they are there! This stock reply has a nice ring to it, implying as it does the indomitable courage and resolve of man and his determination to overcome any obstacle no matter the odds. Yes, the stars are there and from earliest times man has gazed up at them with awe and reverence. Perhaps too, as he has trodden the long and arduous road of history, he has dreamed of a day when great argosies of the sky might take him to them. Men may indeed—almost certainly will—one day go out across immensity to the stars, but it will not be merely because they are there. There will have to be other good and compelling reasons as well.

We have attained a technological level that a century ago would have been regarded as both impossible and fantastic. Moreover, this explosive development has not ceased; it is continuing and accelerating. Unfortunately, it is all too clear that mankind's political and social progress has not kept pace. Over the past several decades democracies, pseudodemocracies, and outright dictatorships have existed side by side. They don't make easy bedfellows; quite the contrary. Advancing technology has exacerbated not just political differences but also their destructive potential, and this trend also seems to be accelerating with every passing year. Humanity, on the other hand, has barely evolved. Therein lies the terrible danger. Conflict between the superpowers on Earth seems, sooner or later, almost inevitable. They may not want it; indeed, knowing the consequences, it is certain they do not. It could and probably will happen by accident. A missileman looking at a radar screen may observe a "blip" that is due to nothing more than a large meteor entering the Earth's atmosphere. He draws the wrong conclusion and, because he has been conditioned to it, presses the fatal button. The result is fearful Armageddon! Alternatively, the persistent feuding among some of the smaller powers—squabbling to which there seems no foreseeable end—could eventually drag the superpowers into a confrontation from which there would be no backing down. No one would win, of course, but that is merely stating the obvious, and Earth would almost certainly be left as a charred, radioactive cinder. Some remnants of its peoples might survive, but it probably would take man a thousand years or more to attain the degree of civilization he so wantonly cast aside. And having regained it, is there any guarantee he would not repeat his mistake? With possibilities such as these in mind it would hardly be surprising if in the future part of humanity, assuming interstellar travel had become feasible, decided to get out while the getting was good and leave the rest to their follies. Surely better by far to experience years of enforced hibernation on a starship, waking up to a new and unspoiled planet, than to die or barely survive in an apocalyptic conflict.

Worldwide nuclear war is not the only threat hanging over our planet. There is another that is more insidious: the increasing overpopulation of our world. At present the population of Earth is increasing by a staggering seventy million per year. The merest modicum of mental arithmetic points very clearly to where this is leading, and only a few seem to comprehend the dangers. This is not just a nebulous threat lying sometime in the future. The process began some time ago and has been steadily gathering momentum. It is not a future threat—it is a present reality. In the poorer parts of the world famine, even now, is virtually endemic. Every year millions of children die of hunger while hundreds of millions are doomed to suffer the effects of malnutrition for the rest of their lives. It has been estimated that every five years the population of the world is increasing by a figure equal to that of the present population of Western Europe. If that doesn't provide food for thought, it certainly should. What renders the situation even more dangerous is the tremendous imbalance between the rich nations and the poor. Many in these poorer nations will understandably try to migrate to the rich, developed countries by any means possible—even by force if necessary. The result could be little wars, the kind that can develop so easily into big ones. But even if the rich nations were willing and able to assimilate the newcomers, the problem would still not be solved. Given a few years, the annual increase in world population would simply overpopulate these countries, too. Migration along these lines is not the answer. It is not even a temporary palliative. Instead of parts of our planet's population living in abject want and poverty, we would all soon have been reduced to the same perilous state. The blunt truth is that this planet of ours does not possess the resources to maintain an ever-growing population. Neither does it have the necessary living room. Though interplanetary travel capability is now being developed, those other worlds and moons that sweep around the Sun in company with our own are simply not suited to life as we know it. Only the merest fraction of the Earth's population could be absorbed by one or two of these sister worlds and even then only under highly artificial conditions. There is no answer to the problem here in our Solar System. Earth's excess "huddled masses" can be accommodated only on suitable planets of some of the nearer stars. Only there on some of the cousin worlds of Earth will the necessary *Lebensraum* be found.

A further valid reason for colonizing distant worlds could very well be fear of persecution and harassment here on Earth. One can envision parallels with the Pilgrims who set sail for the New World in the *Mayflower* to escape religious persecution in England. A possible scenario could be as follows: At some future point in time the nations and races of Earth are somehow welded into one great, all-powerful, all-embracing state. This has seemed highly desirable to a number of sociologists and other "utopians." Since there would be no separate nations, there would be no rivalries and consequently no wars. All would be provided for. At a cursory glance one might be forgiven for seeing merit in this arrangement. However, closer scrutiny reveals many glaring anomalies. It is virtually certain that the many different ethnic groups would continue to feel an identity of their own and that some would wish to secede. It is unlikely that they would be permitted to do so, regardless of their feelings or the merit of their case. A single world government would be desirable only if it reflected the wishes of the great majority of Earth's inhabitants and was run along truly democratic lines. Unfortunately, it is exceedingly difficult to envisage such conditions ever being met. If the nations of Earth are ever united in one state it is more than likely that this will have been brought about by conquest, not consent. The conquered would be held by force. Somewhere would be the supreme power whether vested in one man, one political doctrine, or one self-defined people. Persecution of minorities would be inevitable. And just as inevitably men and women will assert their need for free speech, for freedom within just laws, for freedom of religion, and above all, for freedom of thought. They'll not secure those freedoms on an Earth the like of which has already been so well described in George Orwell's epic *1984*. So if a form of interstellar travel were feasible, a starship named *Mayflower* might leave the "shores" of Earth forever!

So we see that there are at least three looming reasons for terrestrial colonization of far-flung worlds: the threat of nuclear Armageddon on Earth, a hopelessly overcrowded Earth where living standards are rapidly declining and available resources dwindling, and the threat of persecution on a world gone totalitarian.

The second of these reasons has long been seen as the most inevitable, providing the motive for *Homo sapiens* gradually to spread himself throughout the galaxy. But, of course, if many alien civilizations have also spread themselves about, the result could be big, big trouble—interstellar warfare.

But all the reasons still partake of that primeval need to reach the stars because they are there. Being there they represent the ultimate challenge to man.

2

Planning

A COLONIZING ONE-WAY trip to the stars will require meticulous planning if it is to have a reasonable chance of success. If the basic reason is to escape a possible nuclear war on Earth or to leave an overcrowded world, preparations presumably could go ahead without undue interference from the authorities. Indeed, colonial expeditions with a view to relieving pressure on terrestrial resources might well have the blessing and assistance of moderate, sensible Earth governments. Another such trip, the main objective of which is to escape from the clutches of an authoritarian regime, might find itself assisted by that very regime, only too glad to get rid of dissidents. It is more likely, however, that the authorities would seek to foil any such escape, in which case preparations would have to be carried out in secret. This would not be easy, though the call of freedom is a powerful incentive.

As I mentioned in the Preface, any significant planning oversight could not be corrected on a destination world many light-years remote from Earth and our Solar System. The celestial Rubicon would have been irrevocably crossed. There could be no going back—ever! Some of the more obvious planning essentials have been dealt with at length in other books I've mentioned, so we'll only survey them here.

The first is so obvious that it barely deserves mention: There must exist a suitable planet for the colonists to colonize! That means a planet the mass and gravity of which are similar to

Earth's. Slight differences might not matter, and in fact a world where the gravity is slightly less can offer many advantages. But one that causes a 150-pound man to weigh twice as much would hardly be acceptable. His daily problems would be enormous, and carrying such a weight would probably induce early cardiac arrest.

The planet must also possess a breathable atmosphere—one of oxygen diluted by nitrogen in a ratio roughly similar to that of Earth. There also will have to be sufficiently large landmasses. Clearly a world with an appropriate gravity and atmosphere will be useless to colonists from Earth if it is entirely, or almost entirely, covered by water. It is generally accepted that man evolved from creatures that once inhabited the primeval oceans of Earth, but he can hardly be expected to return to a marine environment in an instant. The chosen world will also have to revolve around a central star that is similar in spectral type, temperature, and dimensions to our Sun and be neither too near nor too far from it. In other words, the distance between the two bodies must be just right to provide a suitable ecosphere for human existence. It is as well for us that Earth's orbit lies at the distance from the Sun that it does. Had that distance been a little less, surface conditions would have been too hot for life; a little greater and they would have been too cold. Life on Earth, as we now realize, was a very closely run thing indeed.

It will also be highly desirable for the chosen planet to have an axial tilt of approximately 23½ degrees so that the seasons resemble those on Earth. Presumably colonists could get by without this, but there are limits to what men can endure. One has only to think of Uranus with its axial tilt of 98 degrees and then try to imagine what seasons would be like on that most uninviting world. Finally, a rotational spin of approximately 24 hours will be highly desirable, too, so that the length of the days on the planet don't differ much from the terrestrial days to which we're so accustomed. Some variation from the Earth norm will be acceptable, but a planet that turns once on its axis during a single orbit around its sun, thereby presenting one face to the star forever, won't meet man's needs. Our kind could not be expected to survive long on a world half of which has permanent daylight and the other half permanent night. Perhaps the twilight areas would be habitable, although they would be small in area.

The consequence of all these essential and desirable features is that two important points emerge. First, to gain detailed knowledge of a remote planet calls for a technology much more advanced and sophisticated than ours is now. Second, it is highly improbable that, even when we do possess the requisite technology, we'll be fortunate enough to find a planet so terrestrial in all respects. We'll probably have to settle for one that fulfills only some of these conditions. The essential ones are a breathable atmosphere, a similar gravity, sufficient land area, and appropriate temperature range. The highly desirable are those having similar diurnal rotation period and similar axial tilt.

Presumably terrestrial colonists can eventually become used to days longer or shorter than those to which they have been accustomed on Earth. This applies to seasons as well, but only if these don't involve summers of extreme heat and winters of terrible cold.

It may well be that we will never have the technology to determine the intimate details of planets so far removed from Earth. We should in the not too remote future (at least in respect of the nearer stars) be able to tell which possess planets and which do not. To a prospective colonizing expedition that information is essential. There's no point embarking upon so daunting an enterprise if there are no planets for the colonists to settle. It has been suggested by a few others who've written on this theme that either an instrumented probe or a reconnaissance mission should first be sent to the environs of the star in question. In view of the tremendous distances involved and the consequent time factors it's difficult to take these proposals seriously. There is no easy way around this problem unless some method of "telescoping" space can be achieved. A future colonizing expedition will always be faced with the risk of extinction if there's no suitable planet available when they arrive. Others see the answer in achieving velocities greatly exceeding the velocity of light. They conveniently forget not only the problem of time dilation but also the sheer magnitude of the power sources necessary. To all intents and purposes the velocity of light is a limiting factor, and despite research into the possibility of utilizing tachyons, we must adhere for the time being to the basic precepts of Einstein.

It is very likely that when the day of such an expedition dawns the only definite information available will be that the destination star does possess planets and that in dimensions, age, and spectral class it is similar to our Sun. If on reaching the star the would-be colonists find that none of the planets is even remotely habitable, they'll have two options: a return to cryogenic hibernation after directing their ships toward an alternative star or toward a greatly changed Earth, which they had left those many years before. There is, of course, a drastic and brutal third alternative: a wholesale swallowing of cyanide capsules.

There's another extremely important point that we have, so far, ignored. This omission is quite deliberate for, since it refers to a most vital issue, it deserves a chapter devoted entirely to it. That question, if you've not already guessed, is the attitude of the colonists toward any intelligent species that may already occupy the destination planet. To what extent life proliferates in the galaxy is difficult to estimate, but the possibility of another race of beings found already "in possession" raises difficulties of a practical, moral, and ethical nature. In the planning stages of a colonizing expedition from Earth this is a nettle that will have to be grasped. A policy depending on the particular situation will have to be developed. A temptation to erase the subject from the agenda might prevail but any such concession to the *mañana* syndrome might have serious consequences later. There may be no indigenous intelligent life on the planet, in which case there'll be no problem. There might, on the other hand, exist a very backward people, who could be friendly or hostile—or there may be a very advanced civilization. Each eventuality will require the application of a definite and specific policy. It's an issue that must be addressed from the start.

All adult would-be colonists will have to be highly qualified in the various professions and trades that life in an embryo colony will require. Indeed, it may be easier to list the sorts of occupations that won't be called for, or at least not for many years. Without wishing to be derogatory, it's obvious that a farmer will be of infinitely more use than a stockbroker, a doctor more than a banker, a nurse more than an actress. At the same time there is good reason to include people from other than the scientific, technical, and practical spheres. It would surely be undesirable that the arts of Earth be neglected, for otherwise the colony would be depriving itself of the cultures of our home planet. In any event, a case must be made for the inclusion of musicians, painters, poets, philosophers, and sculptors, if only to pass their arts to generations yet unborn. A new terrestrial civilization on a remote world will be in danger of falling into a new barbarism if the arts are ignored.

Some other writers have suggested that the colonizing expedition should be organized in two disparate waves, the ships of the first preceding the second by two or three years and containing only artisans and experienced supervisors to carry out the necessary land-clearing and building. That's impractical. When we look a little more closely at the proposal the following points emerge:

1. The artisans would, we must suppose, from time to time require the services of health-care professionals, to name but one group. But where would these people be? On the second wave of star ships still perhaps a year or two behind.
2. It might then be argued that a few medical personnel should be included in the initial wave. That seems eminently reasonable, but still there are snags. Landing on a virgin planet would surely also require the practical skills of several other groups of professional people—for example, chemists to check on atmosphere, water, and soil; meteorologists to study the weather patterns; biologists to check for any dangers arising from an alien flora and/or fauna; and geologists to check for dangerous features such as faults, volcanic action, earthquake zones, and other real or potential topographic perils. By the time sufficient numbers of such people had been included, the composition of the first wave probably would not be all that different from that of the second.

Still, the idea of colonists arriving in two waves should not be wholly ruled out if the initial and smaller of these two waves contains a balanced proportion of the representatives of each necessary trade and profession (there would be a strong case here for putting the representatives of the arts in the second wave). There is quite a lot to be said in favor of such a scheme, since the members of the second, more populous wave would be emerging not onto virgin soil but to prepared sites with living quarters, hospitals, and the like ready and waiting.

One of the most important facets in the planning of a colonizing expedition is also being left to another chapter. For want of a better title we will merely refer to this as the genetic aspect. If sensible, practical decisions are not reached and put into effect in this respect, then no matter how effectively other measures have been dealt with, the colony may eventually die out. Human biology must be regarded as of fundamental importance.

3

Indigenous Intelligent Life

FORERUNNERS OF THIS book, as well as several excellent volumes by other writers, have gone at some length into the possibilities of life, especially intelligent life and the forms it might take in other parts of the galaxy. It's not our purpose to do that again here. In this chapter our intention is to ascertain, if possible, the likely attitudes and reactions of a terrestrial colonizing expedition confronted by another race of beings already on, and most likely indigenous to, the planet earmarked for colonization and occupation. We must examine and discuss the practicalities of the problem and the options (if any) open to colonists from Earth as well as the ethics and moral issues involved.

However, since the question of other life in the universe is continually under review, it is worthwhile to update it in the light of current knowledge and belief. This will help us get the matter into proper perspective. Were we able to prove that mankind is alone and unique in the galaxy, the problem would never arise. Although there are those who incline to this belief, the scientific consensus is very much oriented toward the opposing point of view. The question in our present context centers not on whether aliens exist but whether a colonizing expedition from our planet might one day be confronted by intelligent beings from the world of another star. Why, after all, should our own infinitesimal niche in the universe be unique? To accept that it is, is virtually to adopt a pre-Copernican stance. At the same time, we have no reason to believe that the universe is crawling with life. The position as it is presently seen by astronomers can be conveniently summed up as follows:

1. Stars of large mass are better able to contain their rotary forces. They therefore spin rapidly but do not shed material from which planets might condense.
2. Planets are therefore most likely to be found orbiting less massive stars—those having the equivalent of about ten solar masses. This leaves the majority of stars as likely planetary centers.
3. The central star of a planetary system, and hence the original gas cloud from which it formed, must be composed of the same chemical elements found in the Sun and essential for the constitution of planets—oxygen, nitrogen, carbon, magnesium, silicon, iron, and others. This condition is satisfied by the great majority of stars. Thus planets having a composition akin to that of the Earth may be very common.
4. The presence of these elements means that the "building blocks" of life—polysaccharides, nitrogen-heterocycles, and porphyrins—may be very widespread indeed. Thus potential sites appropriate to biological evolution may be extremely numerous.
5. Evolution is a very slow process. Consequently stars of relatively short life-span will have become bloated red giants that consume their planets long before even a moderately complex life-form can evolve. This applies chiefly to stars that are two to three times more massive than the Sun. Habitable planets are therefore most likely to be found orbiting stars with masses less than about one and a half times that of the Sun.
6. It is also considered essential that such stars have a main sequence life of at least 8 billion years.
7. Planets orbiting stars that meet the foregoing requirements must have orbits that remain substantially unperturbed throughout this time.
8. The average temperature on the surface of a planet must fit comfortably between the melting point of ice (0°C) and the boiling point of water (100°C). This factor is dependent on the planet's distance from the central star and on the energy output of that star.

9. Planetary mass must range from .5 to 2.5 times that of Earth's mass, enabling the planet to retain an atmosphere including water.
10. The planet's rate of axial rotation—the length of its day—should be sufficiently fast to minimize the fluctuations between daylight and night temperatures.

Our galaxy, the Milky Way, contains about two hundred billion stars. Of these, some 80 percent fail to meet the requirements outlined above. This leaves 20 percent having masses in the appropriate range—three-quarters to one and a half times that of the Sun. Moreover, these are not members of binary or multiple systems (planets orbiting such systems probably would have very irregular or eccentric orbits). Thus in the Milky Way there may be about 40 billion planetary systems in which life may have developed.

It is presently impossible to detect planetary systems by optical means from the surface of the Earth, though this may eventually be possible from orbiting telescopes. They can, however, be detected by the "wobble" imparted to a star's proper motion through space due to the presence of a very massive body or bodies of planetary dimensions. The "wobble" given to the Sun's path due to Jupiter might, it is believed, be detected from a hypothetical planet in the Alpha Centauri system, some 4.3 light-years distant. Barnard's Star (6 light-years distant) shows this effect, which might be due to the presence of one large planet about one and a half times the mass of Jupiter or alternatively to the presence of three slightly less massive bodies. Bodies having the relatively low masses of Earth, Venus, Mars, and Mercury impart insufficient "wobble" to the Sun's motion to render it detectable at such distances. However, where planets of great mass exist, so also can small ones—for example, in the Solar System we have the large planets Jupiter, Saturn, Uranus, and Neptune as well as the small ones Mercury, Venus, Earth, Mars, and Pluto. Those stars we presently believe to have planets (with the probable exception of 70 Ophiuchi) have masses outside the range three-quarters to one and a half solar masses. Thus 70 Ophiuchi is the only candidate so far for a potentially habitable planetary system. This should not be taken to mean that it has a habitable planet, since we are unaware whether the planet lies at a distance from the star that may render it thermally suitable or whether the planet possesses sufficient mass to retain an atmosphere. The probability of there being a planet satisfying these two conditions is thought to be around 5 percent. On this basis only one planet in 20 of the 40 billion possible planetary systems would be congenial to life. Nevertheless, this would mean about 2 billion habitable planets in our galaxy.

The average distance between stars in the immediate vicinity of the Sun is approximately 5 light-years. If only 1 percent of these have habitable planets, the average distance between such planets would be about 25 light-years. Of the 14 stars nearest to the Solar System likely to possess planetary systems it is reckoned there is a 40 percent chance that at least one of them has a habitable planet. To would-be colonists the key question must be, "Which one?"

The number of planets harboring a species technologically capable of repelling terrestrial colonists is therefore a factor of prime importance. Let us see if we can arrive at a reasonably based estimate. In the first instance it is necessary to estimate the average duration of a technology. This is difficult because the only technology we know of is, of course, our own. The evolution of life on Earth has proceeded for about 3 billion years, but our present technology is less than 100 years old. If we assume that it will last only about 300 years (being in the end "snuffed out" either by nuclear war or exhaustion of the necessary raw materials), we can begin to quantify on the basis that the number of technologies at a given time is as follows:

$$\text{Number of technologies} = \text{number of habitable planets} \times \left(\frac{\text{technological life-span}}{\text{main sequence life of star}}\right)$$

If there are, say, 2 billion habitable planets and the main sequence life of a star is 10 billion years and 300 years that of a technology, we get the following result:

$$\text{No. of technologies} = 2 \times 10^9 \times \frac{(3 \times 10^2 \text{ years})}{(10 \times 10^9 \text{ years})} = \frac{6 \times 10^2}{10} = 60$$

On this estimate only 60 technologies exist in our galaxy at any time. From the would-be colonists' point of view such a conservative estimate must be regarded as extremely favorable, since their chances of being confronted by a more advanced race is most unlikely. However, there is another side to it. Suppose instead we put our own technological duration as 30,000 years, which is roughly a tenth of the time *Homo sapiens* has existed. On this basis we get the following result:

$$\text{No. of technologies} = 2 \times 10^9 \times \frac{(3 \times 10^4 \text{ years})}{(10 \times 10^9 \text{ years})} = 6 \times 10^3 = 6{,}000$$

This figure would give our colonists some food for thought. Nevertheless, 6,000 equal or superior cultures spread throughout the galaxy still represents fairly favorable odds. Now, however, let us assume that a technology could exist for 300,000 years—about the same time that *Homo sapiens* has been in existence. Thus:

$$\text{No. of technologies} = 2 \times 10^9 \times \frac{(3 \times 10^5 \text{ years})}{(10 \times 10^9 \text{ years})} = 6 \times 10^4 = 60{,}000$$

Spread widely throughout the galaxy, 60,000 high-level technologies give more threatening odds, but our colonists still could be quite optimistic.

But we are really only playing a numbers game, since we cannot possibly estimate how long the average technology will last. Neither can we be sure that technologies are spread uniformly throughout the galaxy. Chances are that they are not, and for all we know a cluster of them may occupy that part of space nearest to us. In that case our colonists might easily run into one. The chances of encountering it are enhanced, moreover, because technologies of the standard of our own or better, since they would be indigenous to planets most like our own, are going to exist in the environs of stars like our Sun, the very type of star that a terrestrial colonizing expedition will select as suitable.

It is easy to speak of an Earth expedition, when confronted by a highly advanced civilization and technology, turning away and heading for another star. Unfortunately, this isn't as simple as it sounds. The position of the other star could easily involve a large-angle course change, and in order to achieve this the star ship would have to rid itself of a considerable amount of ballast. This is tied up with a relationship between rate of ejection of mass, rate of energy emission, and stated thrust. This mechanical/physical relationship is complicated and there is little point in going into it in these pages other than to emphasize that proceeding from star to star a number of times simply would become impracticable.

Where, then, do we stand as Earth's first colonizing expedition finally approaches a suitable planet? Obviously if those who believe there is no other intelligent life in the galaxy are correct, there is no problem. The colonists can safely disembark, though safety may be relative in view of the possible existence of animal predators and/or microscopic organisms.

Our colonists, for the reasons already stated, are going to head for a star similar to the Sun knowing that the star in question possesses a planetary retinue and that one or more of the planets might prove suitable. It is also here that an indigenous population could exist—or even a colony of intelligent beings from another star who got there first.

If life similar to our own is the pattern for life in the galaxy, then it is at just these points that potential colonists from Earth are most likely to experience that much-dreaded confrontation. If, however, intelligent life can exist in diverse other forms, then life of that nature, we must assume, is much more likely to be indigenous to planets different from our own and where beings like ourselves could not possibly exist. This being so, colonists from Earth are highly unlikely to come into contact with beings extremely alien to their eyes (and vice versa). This is probably a tough break for all the "sci-fi" buffs who so love the BEM ("Bug-Eyed Monster").

When we consider the human body—its structure and metabolism—it is abundantly clear that we are well suited to our terrestrial environment, just as aliens very different from us could be equally well suited to theirs. However, we must not forget that basic laws of physics and chemistry are common to the entire universe and not to Earth alone. Our bodies are based on the elements carbon, oxygen, hydrogen, and a few others of less importance, and we comprehend the biochemistry that our bodily processes involve. If we try to conceive permutations using other elements we soon come up against serious problems. If we envisage a life-form based on the element silicon rather than carbon we find that although silicon and carbon have a lot in common (especially their valences or combining powers with other elements), the resultant biochemical processes would be totally different if, indeed, they were not precluded altogether. It would seem that carbon is the only element, along with hydrogen, oxygen, and the minor elements, that satisfies biochemical requirements. In short, there is a very strong case for believing that only carbon-based life can exist. Moreover, it really can exist only on planets not too dissimilar to Earth orbiting an essentially Sun-like star. There could be and no doubt are variations, but the degree of difference does not seem likely to be very great. Our colonists are therefore most likely to meet up with life similar to our own at the very points they'll least wish to find it. The full nature of the dilemma is becoming increasingly obvious.

It might be asked, of course, whether life on a planet like ours orbiting a star like our Sun must include intelligent beings among its fauna. It is a very difficult question to answer and certainly not one that can be answered with any assurance. The truth is that we are still by no means certain just how intelligent man arose on Earth. It may have been due to a peculiar set of circumstances unlikely ever to have been repeated elsewhere in the universe. On the other hand, it may simply represent the normal course of evolution, and that seems much more probable.

If all this reasoning (or speculation) is valid, attempts by the human race to colonize Earth-like planets will be fraught with peril. Let's consider some of the situations that may easily arise and try, if possible, to find reasonable courses of action to suit them.

Case 1: The planet has extensive flora and fauna but contains no intelligent beings. In these circumstances the problem does not arise.

Case 2: The planet contains intelligent beings but these are of a very low order, civilizations perhaps equivalent to those of our early Stone Age. In this case there is a comparatively minor problem. The colonists can safely land and begin to take over the planet. The most important question is how they will treat its primitive inhabitants. Though still primitive these people, given a few millennia, might eventually produce a civilization every bit as accomplished as ours. We must not forget that our own ancestors were Stone Age men. We have to be honest and admit that the track record of our own kind in respect to less-advanced peoples on Earth is exceedingly poor. The colonists could decide to be quite ruthless, seeing these "Stone Age" creatures as nothing more than an unmitigated nuisance, perhaps on occasion a dangerous one. They therefore proceed to slaughter them at every opportunity. But if the colonists have quit Earth because of cruelty and oppression then they will merely be emulating the very oppressors from whom they have fled.

Alternatively they might try to treat these simple creatures kindly and humanely and perhaps even gain their trust and friendship. Even though there would be no common language, the cliché is true that action speaks louder than words. Perhaps kindliness is a common language in its own right. Injured or seriously ill members of the race made fit again might well show a childlike devotion to their benefactors as a consequence. They might be taught such elementary skills as clearing land or helping to raise crops. Such creatures might at first tend to be aggressive, distrusting these mysterious strangers who descended from the sky like gods, in which case considerable patience and tolerance will have to be shown by the colonists.

Case 3: The planet contains intelligent beings who have already attained a measure of civilization roughly akin to that of the Aztecs of Mexico or the Incas of Peru. Now the scope of the problem is vastly increased. The descending colonists will, with their sophisticated modern weaponry and technology, have a decided military advantage, and no doubt it will prove relatively easy to subjugate the inhabitants of the planet. As in Case 2 this may be an instance of the oppressed becoming the oppressor. Because of the inevitable language barrier it will be difficult to come to terms with these people, though this handicap ought not to last forever. To white men on Earth the language of the Sioux or the Arapahoe did not prove a permanent obstacle, nor did English to the Indians. When we recall what the conquistadors did to the amazing Inca civilization, our appalling track record shows to the full. An old and cultured civilization was virtually wiped out for no reason other than greed. It would be every bit as much of a crime to destroy an alien civilization light-years from Earth. The best course under the circumstances will surely be to show, if possible, respect for these people and their civilization and to seek friendly coexistence with them. In the medical field there will no doubt be much we could do for them, and if they are basically an agrarian society (as they probably would be) we might be of considerable assistance to them. Equally this people may assist the colonists inasmuch as they will already have cleared land, hunted down dangerous predatory beasts, and erected buildings that can be used for shelter. In that event the colonists will not be descending upon a wild, untamed virgin world. Nevertheless it is very difficult to envisage two such widely differing cultures living amicably with one another, although the attempt should certainly be made. The only alternative if the colonists wish to remain on the planet is continual war or wholesale genocide. Surely there has been enough of that on Earth already.

Case 4: The planet originally had no indigenous intelligent species but has already been occupied and colonized by aliens from the world of another star. The original colonists will not, we must assume, be prepared to permit the people from Earth to wrest the planet from them. If the terrestrials try, a very sanguinary and destructive war may result. Only if the Earth colonists have superior weapons and are prepared to accept a high number of casualties will the aliens be defeated. In the process irreparable damage might be done to the expedition from Earth. And even if our colonizing descendants do defeat the aliens, what then? Will an uneasy peace prevail, broken at frequent intervals by fresh bloody conflict? The terrestrials will find it very difficult if not impossible to keep the alien colonists in more or less permanent subjection, and genocide of such a race will surely be out of the question. To live and let live will be the only real solution, the two groups of colonists dividing the planet into two separate nations living peacefully and helping and trading with each other. Does this sound like too much to expect? It does, and yet there seems no other course.

Case 5: The planet contains an indigenous civilization of a high order either technologically equal or superior to that of the Earth colonists. The answer is very simple; indeed, it could not be simpler. Our colonizing descendants from Earth cannot hope to prevail. If alien colonists ever try to take over Earth we will almost certainly try to repel them with every means at our disposal. In this case the colonists from Earth must go on their way, trying if possible to find an alternative planetary haven.

Case 6: The planet contains no indigenous people and is well suited to our kind. However, a sister planet in the same star system does contain an alien civilization that might not look with equanimity at the prospect of colonists from Earth establishing themselves on a world so close to their own. This is a very tricky situation. Suppose an alien race enters the Solar System and takes over the planet Mars. This is assuming that Martian conditions suit them. What would we on Earth do, and how would we react? We could say in all truth that Mars is of no use to us, so these people from wherever they come are welcome to it. Moreover, if their technology enabled them to cross interstellar space it probably will be as well if we keep our heads down. But we would be troubled, for they might in time cast envious glances at Earth, and we would realize that if they decide to add Earth to their dominions there probably won't be very much we can do to stop them. Our main hope and prayer will be that a species finding Mars and its atmosphere suited to it will not find the conditions on Earth suitable. And so beings indigenous to a planet in a distant star system might simply keep a sensibly low profile, trusting that the people from Earth who take possession of a sister world of their own will find theirs ecologically unacceptable.

These are the most likely possibilities. No clear answers have emerged, only courses of action that under the circumstances seem most appropriate but that will not necessarily be followed.

Only in cases 1 and 5 is the situation clear-cut. Case 1 leads to successful colonization. Case 5 contains all the elements of total disaster. Whereas the colonists from Earth might make provision to seek an alternative planetary home, this, as we have seen, probably won't be a real option. If a continuation of their journey is impossible, then the colonists must either land or perish. It could be as stark and as simple as that.

4

Genetic Aspects

IN AN EARLIER CHAPTER we touched briefly on an aspect of considerable importance to the continuing existence of a terrestrial colony on the world of another sun several light-years from the Solar System. It deserves a chapter of its own, for the importance that genetics has for a single colonizing mission consisting of a few ISTs carrying a limited number of men and women cannot be overemphasized. In all probability, there would be no returning from such a journey, and there might be no reinforcements to follow. If a colony were based on the principle of steady migration with successive waves of ISTs continually bringing batches of new colonists and supplies, then the genetic aspects would have less importance and might hardly enter into the calculations at all. Migration from Earth to a remote star world is such a difficult undertaking that it seems more reasonable to suppose that if it is undertaken in the next few hundred years it will be one desperate venture rather than a planned campaign of space exploration.

It is not impossible that wholesale migration to the stars will eventually come about. But we must consider here what the first and most difficult colonizing missions will be like. There is a parallel to the colonization of the Americas in the fifteenth and sixteenth centuries. The colonists arrived at first in single groups in small ships, landing at different points down the long eastern coastline of what is now the United States. Eventually when the New World had been explored, the number of immigrants grew immensely and large numbers were carried in mighty liners such as the *Olympic* and the *Mauretania,* which shuttled continuously across the Atlantic Ocean. The early pioneers faced many unknown dangers and had been compelled to start from scratch. The first space colonists probably will find themselves in a roughly analogous position. It is also well to bear in mind that the first such missions from Earth probably will select different destination stars. There are a number of stars within a reasonable distance from the Sun that could conceivably possess habitable Earth-like planets. The routes that might be followed and the stars chosen were dealt with at some length in the immediate forerunner to this book, *Where Will We Go When the Sun Dies?*

The would-be colonists are traveling to the environs of a star that all evidence shows is like the Sun and that very possibly has a habitable planet or planets. Their numbers will be limited by the expense of transporting them, and the exact figure will be determined by the resources that people on Earth can afford to apply to the building and equipping of ISTs. Their overriding aim will be to colonize a new planet and create a new terrestrial civilization. Once they have arrived they must, however, successfully reproduce their kind. If by chance the mortality rate on the new planet exceeds and *continues to exceed* the birthrate, then sooner or later that colony must perish.

Consequently the organizers of the mission will have to devote considerable thought to the ratio of men to women. The total number of colonists will obviously depend on the number of ISTs available. They might have little control over the number of ships, but the lower that total the greater the importance of the sex ratio. If sufficient children were not born to the colonists, then *Homo sapiens* on his new world would eventually become extinct. Members of each sex must, therefore, be genetically sound so that once they arrive procreation can begin. Whether terrestrial ethics and conventions with regard to mating, marriage, and the family could continue to be observed under the unique conditions prevailing is a matter we will come to shortly. Related to the genetic requirements is the question of the skills, professions, and trades that would have to be possessed by the colonists. Genetically it is preferable that the future colonists be as young as possible. Unfortunately, the acquisition of professional skills takes many years. To train, for example, as a doctor or a surgeon takes at least seven years, and this makes no allowance for the few more years of practical experience that are highly desirable. This represents one of the more extreme instances, but the fact remains that professional people generally are well into their twenties before they can really be considered competent. Those learning trades do not require quite so much time, but real competence at a particular trade is not acquired overnight. Our future colonists then must be as physically fit, genetically sound, and as young as training and experience in their chosen professions and occupations permit.

So far as the ratio of men to women is concerned there could be much debate, but this writer regards equal numbers of each as the best solution. One or two deaths might occur during the journey, but the ratio would in all probability remain much the same. Casualties could occur on the destination planet, especially in the early days, and it seems reasonable to suppose that these would more likely occur to males. This, of course, is a point in favor of those who would call for a higher proportion of males to females on the first ISTs.

If the colonists stick to terrestrial conventions (or at least those of the present day), then men would marry women rather than simply live with them and in due course the first children of the new world would be born. Just how many children each couple would have is anybody's guess, but since the colonists would be young and fit it seems likely that quite soon, in relative terms, the number of children would equal or exceed that of the original colonists. We must allow for a certain degree of infant mortality, but so long as that was relatively low the population of the colony would expand. Nevertheless, this is based on the premise that all the men marry all the women and start to have families right away. This is unlikely, though from the start the marriage rate among so many young and healthy men and women would be high. This does not mean that all couples would wish to start families at once. Their professional duties could preclude this. A young woman, recently married, who happened to be a doctor would have a strong case for delaying conception. Nevertheless, the attitude of many women would be that in their early to midtwenties they were at the best age for childbearing. The situation would, in all probability, work out quite well, with a steady stream of children being born in the long term rather than a sudden population explosion within a year or so of landing that would represent a heavy drain on resources in that period.

Though expected to procreate in the interests of the colony, there could be a number of couples who, for personal reasons, simply did not desire children. It would probably have been made abundantly clear to all colonists before leaving their native planet that those who married were expected to produce families, but promises made then could easily be broken. It is also possible that some members of the colony would have no desire to marry, though that would not necessarily preclude their adding to the birthrate.

If as a consequence of epidemics or other natural catastrophes a serious imbalance between the sexes was created, this could result in a loss of potential second-generation colonists. It is more probable that such deaths would affect both sexes equally. Still, it is better to consider as many eventualities as possible. Human nature being what it is, if there were an excess number of females they might still conceive by reason of illicit relations with married men. Alternatively, some of the married women could conceive as a result of liaisons with some of the single men. These things happen all the time on our present-day Earth, and there seems no reason to suppose they would not continue to do so on another world a century or two from now. Unfortunately, such events can lead to the most bitter disputes, and it would be in the best interests of the colony that these be kept to a minimum. Situated on another planet 4 or 5 light-years from Earth and utterly on their own, the colonists must, so far as is humanly possible, pull together for the common good.

So far we have restricted ourselves to the concept of love and marriage with the family as a unit. So long as the population continued to increase, this seems still by far the best arrangement. However, if for any reason the sexual imbalance was upset to a considerable extent, the colonists might have to consider alternatives. Despite the other possibilities mentioned above, the most likely reason for an imbalance is a chance natural one—more girls being born than boys, or vice versa. If more girls were born than boys, polygamy might be an answer, but it would be a poor sort of answer from the social point of view, and it would bring with it the dangers of inbreeding.

An imbalance in the relative numbers of each sex might correct itself in time. Besides, in view of research already being pursued today, there seems a very real possibility that by then it will be possible to control the ratio between the sexes by selecting the sex of the unborn child. There is, however, a difficulty in applying this practice to a developing colony. Parents desirous of a son might not take too kindly to a society that decreed the child must be a girl. This smacks strongly of a totalitarian society, of which we have had (and regrettably still have) too many examples on Earth. In the first decades on a new colony planet some form of authoritarian government would almost certainly be mandatory. It could still, however, be one with built-in democratic safeguards. We will deal with this in a later chapter.

Is the possibility of choosing the sex of a future child anywhere in sight? The answer at the present time is a guarded yes. Some researchers in the United States claim that a foolproof technique to achieve this is within sight. Substances known as monoclinal antibodies have been used to identify the proteins on the surface of male sperm in semen samples, thus allowing them to be separated from female-producing sperm. It has been predicted that it is only a matter of time before this method, already envisaged for animals, can be used for humans.

Understandably there is already much debate concerning this within the medical profession, some genetic experts rejecting the concept in its entirety. Emotionally the issue is most difficult. Would sex choice take the joy out of love and childbearing? In a small colony several light-years from Earth such considerations might have to take a secondary place. It has been claimed that here on Earth the development of such a technique would quite likely cause an imbalance between the sexes. This claim is based on a supposed predilection among parents in favor of sons rather than daughters. Whether this supposition is accurate is hard to say without precise statistics—and these do not exist. If the claim is valid it simply means that the technique, if permitted free rein on Earth, would lead to sex imbalance, whereas among the members of a planetary colony it would have precisely the opposite result. The difference, of course, is based on the assumption that on Earth the practice would be uncontrolled, whereas among planetary colonists it would have to be controlled.

Artificial insemination might also have much to commend it. This would involve the use of stored semen taken from males considered too old to join a colonizing mission. This would preferably come from males having very high IQs. Semen banks are now used for artificial insemination. The present ethics (some might be tempted to say prejudices) of our society may retard the practice, but the ability to transmit the genes of men of proven brilliance into young women (also of high IQ) in a terrestrial colony deep in space has much to commend it. If there was ever a considerable surplus of females this might provide the solution for that problem.

It is difficult to predict the genetic aspects of planetary colonization, since we cannot accurately predict what scientific and technological advances will have been made a century or two from now. It is almost certain, however, that when our descendants set out to colonize other worlds they will have made a careful study of the genetic aspects of the venture.

5

Base Camp I

HAVING DEALT WITH THE essential preliminaries, it is now time to think about the actual colony itself. In the beginning this would merely constitute a base camp. From there exploration of the planet will begin, the necessary precursor to colonization of the habitable parts of the new world.

It is clearly going to be of considerable importance which part of the planet and what type of terrain the colonists choose for their first base camp. We are assuming, as we have all along, that the planet is one similar in all important respects to Earth. It is of comparable size and mass, the star it orbits is a near replica of the Sun, the planet is divided between ocean and dry land in roughly the same proportions as Earth (this will tend to give it a terrestrial type of climate, but more regarding that later). The new world also lies at the appropriate distance from the parent star to produce a sufficient but not excessive ambient temperature over much of its surface. Since the colonists can hardly expect to have everything their own way, the length of the planetary day may differ by an hour or two, and its axial tilt may differ by a few degrees.

The reader may be tempted, quite justly, to remark that the first colonists from Earth have been singularly fortunate in having found a world so Earth-like but, as we pointed out earlier, a colonizing expedition could be successful and develop satisfactorily only if it found such a world. That is why it would head for a star of a solar type and age and one known to possess planets. The other vital planetary essentials for the expedition might be able to be determined in advance, depending on the state of technology at the time. At present it seems highly improbable that from a distance of several light-years it would be possible to determine intimate details concerning the physical characteristics of extrasolar planets, but it is best to retain an open mind. Not very long ago many if not most eminent men of science claimed that space travel was a sheer impossibility for all time. As for walking on the surface of the Moon—that was a ridiculous pipe dream. We don't seem to hear from these people today. Recently I was reading a book written during the late nineteenth century that made no bones about the fact that flying was strictly for the birds! At the time of reading it I happened to be comfortably seated in a DC-10 at 35,000 feet over the Atlantic en route to San Francisco, having just enjoyed a very pleasant meal served by an equally pleasant and attractive young stewardess. The writer of that particular book must have been long dead, but I wondered whimsically what he would have thought of his prediction now.

The capacity to determine the physical details of a planet of another sun is not one we possess at the present time, nor are we likely to for the foreseeable future. In the distant future it may be possible, but it is also possible that there are things that, by their very nature, will always be denied us. Unmanned probes could be sent ahead of the expedition to survey the planet, but it is unlikely that colonists would be willing to wait for their return. Colonizing expeditions may always have to face the unpleasant fact that on reaching the star system of their choice there may either be no terrestrial-type planets, or if there are, they may be too near or too far from the star to render them habitable.

Assuming colonists reach the star system safely and find a habitable, Earth-like planet, we would regard it as improbable that they would descend to its surface immediately despite an understandable temptation to do so. It seems more likely that they would begin to orbit the planet using the ship's sensors to ascertain something of the planet's topography, climate, geology, and biology. Having assimilated as much knowledge as possible, the colonists would then have to decide which part of the planet's surface would make the best initial base camp. The huge ISTs that have borne them from Earth would not be brought down onto the planet's surface. Instead, with the ships in a suitable parking orbit, a fleet of small shuttle craft (which we will discuss in the next chapter) would ferry down personnel and essential supplies.

As far as a site for the base is concerned, certain areas can be ruled out at once—oceans and seas; polar regions; equatorial regions; dense, selva-type jungles; arid deserts; and rugged, mountainous terrain. Climatic factors would be of the utmost importance, and a coastal region in a temperate zone would probably be ideal. However, it could have its disadvantages, and the one that at once springs to mind is the risk of tsunami-type "tidal" waves. These are generally created by earth tremors or volcanic eruptions on the seabed and may travel for considerable distances, showing their devastating potential only when they hurl themselves on an exposed shore. Even ordinary tides, given the right conditions, can do considerable damage, and at times the combined energies of waves, tides, and high winds can have a devastating effect. A classic case is that of January 1953, when a high spring tide, storm waves, and winds of 115 miles per hour raised the level of the North Sea 10 feet higher than usual. This "surge" in the sea caused very extensive flooding in eastern England, while in the low-lying Netherlands 4.3 percent of the country was inundated, about 30,000 houses destroyed or damaged, and 1,800 people killed.

Tides are alternate rises and falls of the ocean's surface brought about chiefly by the gravitational pull of the Moon and the Sun. The tidal effect of the Sun is only 46.6 percent that of the Moon. In our context this is interesting. The hypothetical planet we are considering might or might not have a moon. If it does and that moon is larger and/or closer to the planet than the Moon is to Earth, then the tides it creates will be greater than those on Earth's oceans, for which our Moon is so largely responsible. Should the planet have no moon or a very small one, then tides would be less than here on Earth. Tides are affected by the shapes of ocean basins and landmasses, and that would apply on another planet as well. On the oceans and seas of Earth waves seldom exceed 39 feet in height, although in 1933 one 112 feet high was observed. It should be clear, therefore, that the colonists' base camp if located in a coastal region could be in peril from tidal action.

It would also be important to acquire as thorough a knowledge of the planet's geology as possible. We have seen how seismic and volcanic action can create devastating tsunamis, but this by no means represents the only geological risk to a base camp. A base established over a major fault system that happened to be tectonically active could be at serious risk. The type of fault system we chiefly have in mind is something comparable to the notorious San Andreas fault that traverses California in a roughly north-to-south direction and puts at risk one of the loveliest cities in the world—San Francisco. Faults are not always obvious, but geologists, by virtue of their training, can generally detect the presence of a fault by a close study of a region's topography or by use of seismometers revealing small tremors of about 1 or 2 on the Richter scale. It is therefore possible that the site of the base camp might have to be moved hurriedly even before it was properly established.

It seems most unlikely that the base camp would be sited near a volcano no matter how long extinct it appeared to be. Geologists are extremely wary about using the word "extinct" for a volcano. "Dormant" is a much more applicable term—and a much safer one. Some of the world's worst volcanic disasters have occurred in the vicinity of "extinct" volcanoes. The classic case is Vesuvius, which, with little or no warning, one day in A.D. 79 blew off most of its summit and obliterated the thriving little Roman towns of Pompeii and Herculaneum. The volcano had long been thought extinct, but it was only dormant. Dormancy can last for hundreds and in some cases even thousands of years. A fairly recent example is Mount St. Helens in the state of Washington; the volcano "blew its top" at 8:32 A.M. on Sunday, May 18, 1980. Mount St. Helens was certainly known to be volcanic but since 1857 had shown no signs of activity. For 123 years it had lain in geological slumber, and most residents of the area assumed it would slumber on eternally. Clearly a colonists' base camp in the vicinity of any mountain that possesses volcanic features is out. It is reasonably certain that so Earth-like a planet would show volcanicity much as Earth does. It is now known that volcanoes are the result of tectonic shift or continental drift, as it is also called.

Neither would it be a good plan to select a site that suffered from seasonal winds like the sirocco or mistral, and most certainly the probability of hurricane-type or typhoon-type storms would have to be considered by the colony's meteorologists. Hurricanes and typhoons form over warm oceans and comprise fast, spiraling winds that may reach 150 to 200 miles per hour. The calm center, generally termed the "eye," contains warm, subsiding air. The "eye" varies from 4 to 30 miles in diameter, while the hurricane itself may have a diameter of 300 miles. The warmth of the air within the "eye" contributes to low air pressure at the surface. Warm, moist air spirals upward around the "eye." Cumulonimbus clouds, often called "thunderheads," are created by condensation. Latent heat is thereby released, and this further increases the upward spiral of air. Hurricanes prove especially destructive along coastlines where storm waves and torrential rain cause destructive flooding. The southeast coast of the United States is particularly prone to this type of storm, September being one of the worst months.

As well as determining the probability of hurricanes, the meteorologists would have to consider what risk there was, if any, of tornadoes (popularly known as "twisters" because of their unique and peculiar configuration). These violent whirlwinds cover a much smaller area than do hurricanes and typhoons. A tornado forms when a downward growth starts from a cumulonimbus cloud. When this funnel-shaped extension of the cloud reaches the ground, it may be between 160 and 1,600 feet wide. It crosses land at velocities up to 40 miles per hour but generally dies out after about 20 miles. Some, however, are known to have traveled as far as 300 miles. Hundreds of these peculiar storms strike the United States each year, usually in the states of the Midwest. They are extremely destructive.

One very definite advantage of a coastal site for the base would be the presence of seawater. Though ideally suited in a number of ways, the site could easily be entirely devoid of fresh water other than what could be collected from rain. Seawater contains a high proportion of dissolved salts and this renders it unsuitable for general purposes, but this seawater could fairly easily be converted into pure, fresh water by simple distillation equipment brought from Earth. The end product is distilled water. It is perfectly drinkable in spite of its rather "flat" taste. The question of water supply is important, and if the site chosen could provide natural pure water in abundance, the early months of the colony's existence would be eased. The proximity of a river, stream, or lake, assuming the water contained no dissolved harmful substances, would be ideal. Failing this, a supply of "ground water" would suffice. "Ground water" commences its career as rainwater. This enters permeable rocks through the zone of intermittent saturation. This is basically a layer that may retain ground water after heavy or continual rain but that soon dries up. Beneath this layer is a rock zone in which the pores and crevices are filled with water. This zone is the zone of saturation. It usually begins within about 100 feet of the surface and extends downward until eventually it reaches impermeable rock, through which water cannot percolate. This impermeable rock layer located below the water-holding layer (or aquifer) is a ground water dam. The top of the saturated zone is the water table. It is far from being a level surface. Beneath plains it is fairly close to the surface, but under hills it is often arched. During the year it varies in level depending on the amount of rainfall. In certain places the water table intersects with the surface, creating such features as oases, swamps, lakes, and springs. Springs are seepages of water, sometimes with considerable force behind them, that may occur along the base of a hillside or in a valley in hills. They are found where the water table or an aquifer appears at the surface or where the aquifer is blocked by impermeable rock. Springwater enjoys the considerable advantage of nearly always being fresh and clean, having passed through the pores of rocks such as sandstone, which filters out impurities.

Artesian wells are another source of clean, pure, and potable water. To secure water in this way wells must be drilled down as far as the permanent water table. Water is forced to the surface by hydrostatic pressure. It may seem that we have gone into the subject of water at more length than the subject demands, but the inescapable fact remains that human beings cannot exist for long without an abundant supply of pure, fresh water. To the colonists from Earth in their first base camp on a remote planet, a water supply must have a very high priority. We can hardly expect one of the ISTs in the small fleet to have served as a super water-tanker and, even if one had, the amount if carried would not relate to the needs of the colonists. As we said earlier, distilled seawater would serve, but this requires the application of heat—energy, though the ship's atomic reactors could presumably supply this. However, any reader who has had the misfortune to drink distilled water will not have relished the experience, unless he or she was very thirsty.

Here then is an essential first task for the geologists among the colonists: to try to select a site for Base Camp I in which there is a strong possibility of abundant, indigenous supplies of fresh water. It would then be up to the chemists to make sure it contained no harmful dissolved salts. The fact that water is fresh does not mean it is entirely free from dissolved salts. The bacteriologists would then need to ensure that it was free from dangerous bacteria.

In the selection of a landing site and initial base it would also be wise to examine the flora and fauna there. There could be plants that were highly poisonous to human beings. I don't wish to suggest that the colonists would be so foolish as to eat berries or fruits that were unknown to them. There could, however, be plants having barbs or hairs that on accidental contact with the skin could lead to irritation or even sickness and death. Here on our own planet plants of this nature are known. A classic example is the stinging nettle. It grows in temperate regions and belongs to the family Urticaceae. There are 35 species of Urtica, all of which have bristlelike stinging hairs that are actually long, hollow cells. The tips of these are toughened with silica and are easily broken off. When the plant is touched the hairs penetrate the skin like hypodermic needles, the tips are detached, and the poison within the cells is released. Its effect on people varies, and generally no great harm is done. Some people, however, are highly allergic to the poison. Who can tell what the effect of unknown species on an alien planet might be? All that has been said of plants will probably be even more true of insects and reptiles. It is easy to compile a list of such pests inhabiting our own planet. For instance, in the case of insects there is the common wasp. Its sting is painful and if inflicted in the region of the eye can be very dangerous. Then there are mosquitoes, which it would be an understatement to call harmful. Not only do the little creatures suck our blood, they also transmit such diseases as malaria, yellow fever, and elephantiasis while doing so. A full list of such pests would be quite lengthy, especially if we included some of the more exotic forms of tropical insect life. We cannot predict how evolution will have proceeded even on the most Earth-like planets, but it is a fairly safe bet that it will have produced equally unpleasant creatures. Reptiles cannot be ignored either, and among these snakes come first to mind. On Earth dangerous snakes fall roughly into two categories—the poisonous and the constrictors—and 2,800 species of snakes are known. Among the poisonous variety we can include the rattlesnake and the cobra. The constrictors are generally large. The boa can be up to 12 feet long, while the anaconda of the South American rain forests can occasionally exceed this. Snakes and other reptiles tend to be more common in tropical than in temperate regions, so if reptile evolution has followed a similar course on other Earth-like planets this would be an additional reason for establishing a first base camp in temperate rather than in tropical regions.

Colonists from Earth would also be required to exercise the greatest caution with respect to bacteria and viruses. Readers of H. G. Wells' legendary *War of the Worlds* will no doubt recall that when all Earth's military strength had proved useless against the invading Martians the people of Earth were eventually saved by terrestrial bacteria. Over thousands of years they had developed an inbred resistance to them, whereas the Martians had not and soon died as a consequence. Conversely, Earth people might have no powers of resistance against the indigenous bacteria of an alien world. Bacteria and viruses are the smallest forms of life on Earth. Though simple creatures they are not primitive and have a unique capacity for survival in the most inhospitable of places. They are also well able to adapt to new conditions, a quality that places them among the most successful and advanced of living things. Bacteria are almost invariably associated with disease, death, and decay. They return essential nutrients to the soil and produce complex food that can be used by other living creatures. They are therefore essential to the balance of nature.

Viruses are quite different and have been described as "living chemicals," a fairly appropriate term in the circumstances. They neither feed, breathe, grow, nor move, and they are never found living free in nature. Yet once they enter a living cell they can take control of it, subverting it from its normal activities to the production of new viruses. Unfortunately, all living cells are susceptible to attack by viruses and, somewhat ironically, even bacteria succumb to a special virus known as bacteriophage. A particular virus will attack only one organism or part of that organism. They destroy the cells they attack, causing the release of new viruses. These cause diseases such as yellow fever, poliomyelitis, influenza, smallpox, and the common cold. Can we be certain that similar unpleasant things of a viral nature would not be present on an alien world? The answer is stark and simple: We cannot. Bacteria and viruses and the menace they represent should not, of course, be seen only in the context of the colonists' base camp. Even on the shuttles flying between the planet's surface and the huge ISTs in their parking orbits they could prove a serious danger. Once any of the colonists breathed the planet's atmosphere they could be at risk, and any diseases they developed as a consequence could be transmitted to those in the ISTs who had not yet gone down to the surface of the planet. Bacteria and viruses represent a considerable hazard.

In view of all these dangers, it would be almost mandatory to dispatch two or three shuttle craft to the intended base site first. Among the personnel aboard would be geologists, chemists, bacteriologists, biologists, and botanists, some armed military personnel for protection against possible large predators, as well as all the testing and sampling equipment. These people would need to have protective clothing, life-support systems and their own air supply until known dangers had been identified. The shuttles would need to be equipped with airlocks so that air from the planet's atmosphere would not be carried up to the ISTs until it had been ascertained that it contained no harmful bacteria or viruses. If it did, determined efforts would be made in the laboratories of the ISTs to produce appropriate antibiotics. This might be neither an easy nor a quickly accomplished task. It would be the cruelest of ironies if having survived the dangers of interstellar space and perhaps several years of induced hibernation, the colonists were to be wiped out or seriously depleted in number by one of or a combination of the perils with which we have just dealt.

We must remember that the chosen planet, although Earth-like in so many respects, might still be less than ideal. Since there is no other and there can be no going back, these space pioneers must make do with what they have. To proceed to another star system (assuming this is technically and logistically feasible) might amount to discarding substance for shadow. The next planet might lie a few million miles farther out from its parent star, in which case its surface would be colder. Its polar ice caps would cover a greater area, and its equatorial belt would be the area most appropriate for living in. If the converse were true and the planet lay a few million miles closer to its star than Earth does, then the equatorial regions would have to be avoided because of excessively high temperatures, and suitably temperate conditions would be found nearer to the polar regions. And those, of course, would be the best possible results of proceeding to another solar system; at the worst, life would be insupportable.

It may seem that our colonists in seeking a suitable site for the all-important Base Camp I are bent on achieving perfection—in other words, the impossible. This may well be true and they would be fortunate to discover one that fitted all their requirements with no corresponding disadvantages. But our colonists will be realists. When they leave Earth on their way to a new star world they will have assessed not only the perils of the journey but also the abundant perils that await them on even the most suitable world. Thus many, or perhaps just a few, of the dangers and disadvantages discussed in this chapter will simply have to be accepted. Those courageous men and women who won through and created a new, dynamic nation on the other side of the Atlantic two and three centuries ago faced perils and discomforts, too. However, the human race is at its best when the chips are down.

6

Shuttle Craft

BEFORE WE DEAL WITH further colonization we should devote a little attention to the ubiquitous shuttle craft and see if we can suggest something of their possible forms. We are still perhaps centuries ahead of the day when *Homo sapiens* will burst the bonds of the Solar System, and it is extremely difficult to reach hard and fast conclusions in such matters. Indeed, this could be said of the entire concept of a colonizing mission from Earth to a suitable world of another star. All we can do is peer myopically through the mists of time.

The probable forms of the star ships, the ISTs, have been dealt with in some of this book's forebears and at the present time there is little or nothing we can profitably add. They would certainly have to be of gargantuan proportions, but those used for "hibernation" travel would in all probability be somewhat smaller than the type necessary for "generation" travel. With their occupants in the trancelike sleep of suspended animation, less space would be required for habitation and none for recreation. Nevertheless, so much essential equipment and supplies would have to be carried that the ships still would assume massive proportions.

It may be feasible and might appear highly desirable to bring the ISTs down from their parking orbits to the surface of the planet where they could provide excellent shelter and protection during the colonists' early days. There is no doubt, however, that this would prove a most hazardous undertaking. The great ships could easily suffer total and spectacular destruction along with their remaining occupants and equipment during the descent. Better by far to avoid the risk of so great a catastrophe by keeping the ships in their parking orbit around the planet and continuing to use various types of shuttle craft to ferry down personnel and supplies.

Shuttle craft would almost certainly differ in form and size depending on the role they were designed to play. Five different types would seem to be necessary, which we have designated as follows:

- Type 1A (exploratory)
- Type 2 (personnel)
- Type 3 (freight)
- Type 4A (freight and bulk haulage)
- Type 4 (personnel)

SHUTTLE CRAFT TYPE 1A (EXPLORATORY)

In the preceding chapter we suggested that those personnel first down to the planet's surface would almost certainly need to be specialists bent on testing out suitable sites for Base Camp I. From the ISTs in their respective parking orbits it would be possible to scan almost all of the planet's surface, and several potential landing sites probably would suggest themselves. The meteorological aspects of the planet would almost certainly be gauged better from an IST, enabling the meteorologists to build up a fairly accurate picture of the planetary weather system and then, in the light of this information, to assess the dangers of hurricanes, typhoons, tornados, and other dangerous climatic phenomena in respect to possible landing sites. But accurate geological, biological, and bacteriological data are much more likely to be gained on the planet's surface.

Since these first descents to the planet's surface are going to be exploratory, something much more versatile than a mere space ferry would be required. The colonists would need a craft with a twofold function. It would have reaction power units to propel it between its IST and the planet's surface, as well as additional aerodynamic features enabling it to fly through the planetary atmosphere like a conventional aircraft. In other words, it must be a hybrid—a combination of space vehicle and aircraft. Inevitably we think of the highly successful *Columbia* developed by NASA. No doubt the flying shuttle we are envisaging will be much more versatile and sophisticated, but after a century or two this is to be expected. It would be sufficiently large to carry several persons plus a considerable amount of equipment and test gear. There would be airlocks to prevent ingress of viruses. As with all types of shuttle, retrorockets and efficient heat shielding would be necessary to protect it from the frictional effects of the planet's atmosphere. Once safely within that atmosphere the craft would assume a horizontal attitude and commence to fly as a more or less conventional aircraft, though powered now with an internal-combustion engine to conserve its limited stock of rocket fuel for its brief sojourns in space. *Columbia* has been provided with a fairly conventional landing gear since NASA has the use of a natural dry lakebed in California. On the planet there might be no such feature. The winged shuttle would therefore require VTO (vertical takeoff) features. This would enable the shuttle to touch down at almost any point on the continental areas of the planet. A form of air-cushion undercarriage could be provided. In the still unknown climatic features on the planet, pilots of these machines would need to be wary of peculiar cloud formations, unexpected crosswinds, and downdrafts.

SHUTTLE CRAFT TYPE 2 (PERSONNEL)

Once a suitable landing site had been selected and approved by the exploratory experts, a start could be made in bringing down a first batch of personnel from the ISTs. This group would also be specialists, consisting almost wholly of skilled technicians and artisans. Its members would be responsible for constructing the base. On Earth, construction sites (especially during wet weather) have all the appearance of chaos from which it seems order never will emerge. On an Earth-type alien planet the same probably would be true. These particular shuttles would be relatively small, though larger than the exploratory Type 1A shuttles just discussed. The Type 2 shuttles would carry about 20 persons each but would need to be sufficiently large to serve also as temporary living quarters for the personnel they carried. Adequate accommodations might take some time to construct, though if structures were prefabricated only a small degree of site preparation would be necessary.

SHUTTLE CRAFT TYPE 3 (FREIGHT)

The technicians ferried down to prepare the site, construct the buildings, and provide essential services such as water and electrical supply and sewage disposal could hardly be expected to achieve these things with their bare hands and a few shovels. A great deal of heavy equipment will be essential, ranging from earth-movers, tractors, mechanical shovels, bulldozers, and tools (both powered and manual) to prefabricated buildings, pipes for water and sewage, electric generators, explosives, fuel, and food. The list seems endless. It goes without saying then that the Type 3 freighters would have to be of considerable dimensions and equipped with adequate loading ramps for the vehicles. A fleet of them would be necessary and there would be a considerable amount of shuttling as fresh materials and supplies were required by the growing base. As with the Type 2s these shuttles would have no aerodynamic potential. Some of them would need to be equipped as mobile workshops to facilitate constructional work.

SHUTTLE CRAFT TYPE 4A (FREIGHT AND BULK HAULAGE)

As with Type 1A, the suffix "A" stands for aerodynamic. These would simply be very large versions of the Type 1A shuttle craft. It is perfectly conceivable that raw materials indigenous to the planet could be available and would prove useful. There is, for example, timber (we can assume that an Earth-like planet would have trees of one sort or another); minerals that could be extracted from the ground for later smelting, including copper, zinc, and tin; and hard rocks for subsequent crushing to form the basis of roads and pathways. If pure, fresh water could be obtained only at some distance from the growing base, some Type 4A craft could serve as convenient aerial tankers until such time as an adequate supply of water could be piped to the base by conventional means.

SHUTTLE CRAFT TYPE 5 (PERSONNEL)

Shuttles of this type would simply be greatly enlarged versions of the Type 2 but capable of transferring colonists from the ISTs in much larger numbers—say, 100 at a time. Some parts of the base would certainly be completed ahead of others, and therefore the transfer of colonists would be a carefully phased operation. The first large batch probably would arrive once such facilities as hospitals, meeting halls, stores, and some living accommodations were complete. In the case of hospitals this would imply they were adequately equipped and staffed. Utilities such as water, electricity, and sewage would need to be functional. Until then colonists would be better served on the ISTs, where all necessities would be available. Supplies of water on the ships might be limited, but more could be brought up by means of Type 4A tankers.

7

Communications

WITH BASE CAMP I AS A suitable "beachhead," colonists from Earth would almost certainly wish to begin exploring their new planetary home. They would desire to pinpoint sources of minerals, so it would be essential to achieve a thorough knowledge of the planet's geology. The colonists, after a period of consolidation, would branch out from their initial base camp and develop others. It would be in their interests to discover areas appropriate for growing food. The question of food will be dealt with in more detail in a subsequent chapter.

Exploration of the planetary surface at once introduces the question of communication between exploration parties and their base and eventually, as the number of bases grew, contact among these.

Exploration parties could conveniently use shuttle crafts Types 1A and 4A, as described in Chapter 6. They could also, as we do on Earth, use conventional aircraft (fixed-wing and/or helicopters) or tracked vehicles. In this respect the Type 4A might prove the most advantageous, since their greater dimensions and carrying capacity would permit their carrying aircraft and/or vehicles. Fixed-wing aircraft would have to be of the VTO (vertical takeoff) variety since there would be no prepared runways and no guarantee of suitable strips. Even that much-maligned aircraft of an earlier age might serve a useful purpose—the airship or dirigible, so long as supplies of helium lasted (hydrogen would achieve better lift, but airships obtaining their lift from this gas have a nasty propensity to burn, as past experience has taught us). A maximum amount of provisions, supplies, and fuel could be packed into the 4A's, with separate quarters for the men and women of the exploration parties. These craft also could shuttle up to the orbiting ISTs for further supplies, equipment, or medical aid.

Communication must clearly be carried on by radio. This does not constitute any particular problem as long as the ionospheric layers in the planet's upper atmosphere are similar to those of Earth. For those readers unfamiliar with the principles of radio communication it may prove useful to explain the basic essentials as briefly as possible.

When electrons oscillate (move rapidly backward and forward) some of their energy is converted into electromagnetic radiation. The frequency or rate of oscillation must be of a fairly high order to produce radio waves of sufficient intensity. These waves travel through space with the velocity of light, which is slightly in excess of 186,000 miles (300,000 kilometers) per second. When such a wave meets an antenna some of the wave's energy is transferred to free electrons in the antenna, causing them to flow as an alternating current having the same frequency as that of the radio wave. The antenna is connected to a radio receiver, which can be tuned to the desired frequency and the signals amplified.

The transmission of sounds is achieved by modulating the carrier wave (of a specific frequency) generated in the transmitter by the relatively low audio signal of the speech or music. Thus instead of a carrier wave of the same amplitude we now have one in which the amplitude varies in accordance with the fluctuations in the amplitude of the audio signal. This mode of transmission is termed amplitude modulation or, more simply, AM. Alternatively, the frequency and not the amplitude of the carrier wave can be modulated by the audio signal. This is known as frequency modulation or FM. The receiver is able to eliminate the high-frequency carrier wave since it has served its purpose, leaving only waves of the same frequency as that of the original modulating sound. These are then amplified and fed into headphones or a loudspeaker, both of which can convert electrical vibrations back into normal low-frequency sound waves. Relatively low-fre-

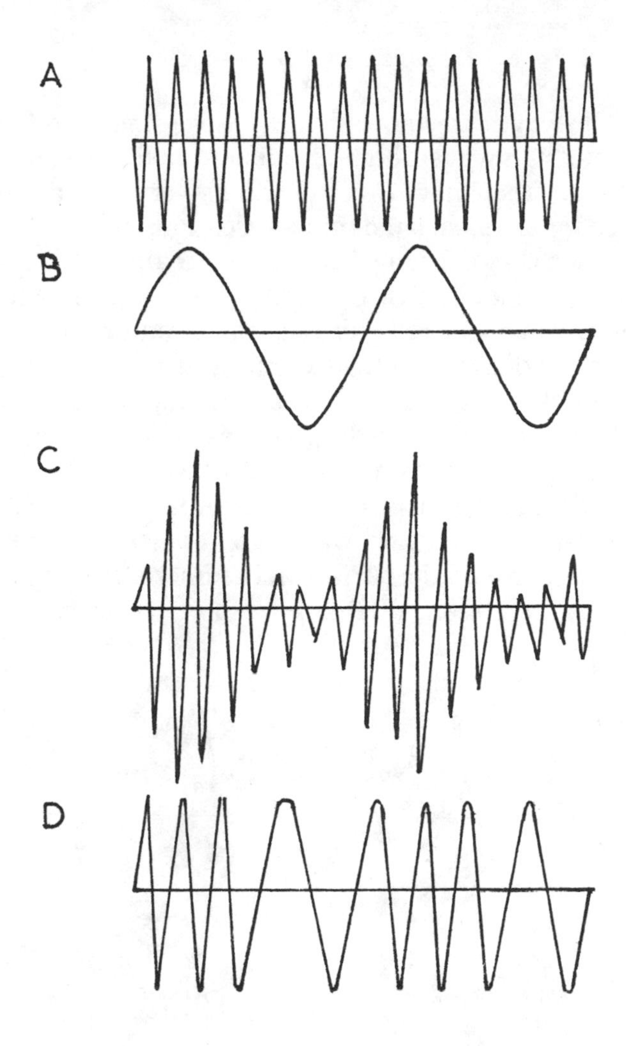

FIG. 1.

Radio waves have high frequencies (A) whereas sound waves have very much lower frequencies (B). In order to transmit sound by radio it is necessary to superimpose the sound or audio frequency onto a radio wave known as the carrier wave. This is known as amplitude modulation or AM. The audio wave modifies the amplitude of the carrier wave to produce an envelope of varying amplitude (C) corresponding to the sound or audio waves. In frequency modulation or FM, the carrier amplitude remains constant, the wave's frequency being increased or reduced to produce a frequency replica of the sound (D).

quency radio waves are known as long waves, medium-frequency waves as medium waves, high-frequency waves as short waves, and very-high-frequency waves (VHF) as very short waves. All can be used for broadcast and communication, though the distances over which they can travel vary considerably. This is due to a layer or layers of ionized gases in the Earth's upper atmosphere known as the ionosphere. This, acting in conjunction with the curved surface of the Earth, serves as a form of "wave guide." This has the effect of bending the path of long waves around the Earth. The wave path of medium waves is not bent so acutely, and for this reason they cannot normally be received more than a few hundred miles from the transmitter. All radio waves travel in straight lines, but since short waves are reflected not only by the ionosphere but also by the surface of the Earth, they can be used for worldwide reception. VHF or very short waves pass straight

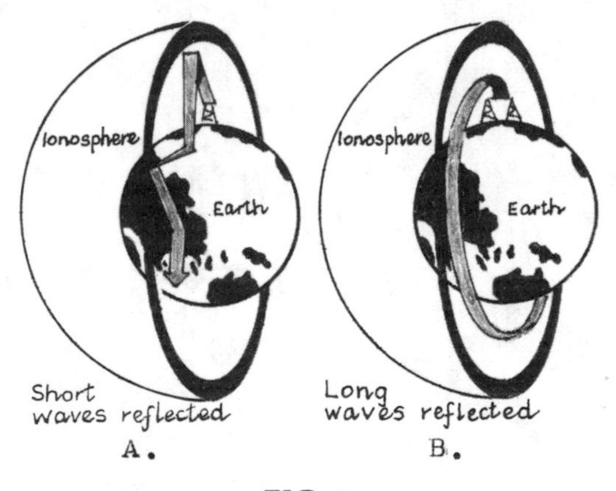

FIG. 2.

The ionosphere and the curved surface of the Earth act together as a form of "wave guide" that bends the path of long radio waves (B) around the Earth. The path of medium waves is not as bent and thus they cannot normally be received more than a few hundred kilometers from the transmitter. While radio waves travel in straight lines in free space, short waves are reflected by the ionosphere and also by the Earth's surface (A). They can therefore be used for worldwide radio communication. Even shorter waves (VHF and UHF) pass through the ionosphere and so are used for space communications.

through the ionosphere, which renders them an excellent medium for communication between Earth and space vehicles or between the Earth and the Moon as long as they are accurately aimed. This necessitates a special type of beam antenna.

With all this in mind let us return now to the colonists and communication between or among bases, exploration parties, and the ISTs in orbit around the planet. So far as the last is concerned, VHF (very short waves) or UHF (ultrashort waves) would be used. Communication between or among bases, if they were no more than a few hundred miles apart, would be quite well served by efficient high-power medium wave transmitters. Where the distances between them were longer, short waves would need to be used. The exploration missions, since they might at times be far from any base, would also use the shortwave bands. Between individual members of an exploration mission who had become separated, hand-held, battery-powered VHF/FM transceivers would prove ideal so long as high ground did not intervene.

So far so good, but we are arbitrarily assuming that the planet has ionospheric layers in its upper atmosphere. But suppose that the planet, though essentially Earth-like, differs in one notable respect—it lacks an ionosphere. Then long waves, medium waves, and short waves would then act much like VHF. They would pass straight up into space, and apart from the transmitters' ground waves (which generally travel only a few miles), radio would have become virtually useless on the surface of the planet except over short distances using VHF.

Let us assume, therefore, that the new planet lacks ionospheric reflecting layers. This being so, what is the answer as far as radio communication is concerned? There are two possibilities. The orbiting ISTs could receive all radio signals from the planet's surface—not just VHF and UHF—and by means of suitable antennas reflect these back to the planet's surface in the required direction. However, antenna restrictions might render this difficult in the case of long and medium waves. Optionally or additionally, a number of satellite relay stations placed in orbit from the ISTs could also serve this purpose. Such relay stations are already in orbit around Earth and are used to enhance the range of VHF stations. Amateur radio operators having only VHF equipment find these make possible Earth-to-Earth contacts of thousands of miles. These satellite relay stations are known as "Oscars," but the VHF station desirous of using them must also possess a parabolic "dish" antenna capable of following the course of the satellite across the sky. A planet as Earth-like as the one we are considering would be fairly certain to possess an ionosphere. The planet has an atmosphere of oxygen/nitrogen similar to Earth's and breathable by the colonists. Let us consider Earth again. The ionosphere is an integral part of the terrestrial atmosphere. The Sun emits radiation of several kinds: About 48 percent is infrared radiation, roughly 43 percent is radiation visible to the naked eye, while the remaining 9 percent comprises ultraviolet radiation and X rays. The total radiation emitted by the Sun is not received at the Earth's surface for, as the radiation passes through the atmosphere, the radiation is subject to absorption and scattering as well as reflection by clouds. The radiations having the shortest wavelengths—ultraviolet and X rays—are absorbed by gases in the upper atmosphere, leading to certain photochemical reactions (which is as well for us, since these forms of radiation over a short period of time can prove extremely harmful). In absorbing ultraviolet and X rays the molecules and atoms of these gases may lose electrons, becoming positively charged ions in the process. The region of the atmosphere in which the concentration of ions and electrons is greatest (60 to 300 kilometers above the surface) is the ionosphere.

A similar planet with a like atmosphere orbiting a Sun-type star would certainly emit the same types of radiation as the Sun, and the planet's atmosphere would treat them as Earth's atmosphere treats the Sun's radiation. There should therefore be no problem for the colonists in respect to radio transmission and reception other than occasional difficulties due to unusual atmospheric effects that, from time to time, plague radio operators on Earth. Sometimes, though, these disturbances actually help increase the range of radio waves. Typical of this is the aurora borealis or "Northern Lights," which at times enable low-power VHF transmissions to be received at distances greatly in excess of what normally would be anticipated.

By the time a colonization mission from Earth has reached a suitable planet of another star, radio will doubtless have become much more sophisticated than at present. To what extent remains debatable. Already in less than a century radio has gone from spark transmitters of very limited range, capable only of using Morse code, to today's fully transistorized equipment the signals from which can reach the Moon and far beyond.

In this context we should also think about television. Unlike the telegraph, telephone, and radio, television was developed purely as an entertainment medium. Today, however, it has a host of other applications, not least among these being remote surveillance. It might be advantageous for a base camp to be able to see as well as hear how an exploration party was faring. At the present time television equipment, especially on the transmission side, tends to be bulky, though not to the extent it was 30 years ago. Whereas the base with all its facilities could use elaborate and powerful equipment, the explorers would require something akin to the hand-held radio transceivers of today, the "walkie-talkies." There seems little reason to doubt that suitable equipment will have been developed long before man is ready to set out for the stars.

In order that two radio signals do not interfere with one another, the difference between their respective frequencies must not be less than the highest frequency of the signal being transmitted. Television signals use very high frequencies, so that a bandwidth above 5 MHz is regarded as the practical minimum. Since this is equal to the radio space occupied by 500 voice channels, the entire short-wave radio band could accommodate only 5 television channels. For this reason television signals are transmitted on VHF and UHF bands, which can accommodate up to 80 channels. Since frequencies of this order tend to pierce the ionosphere, the range from point to point over the surface of the Earth is extremely limited. This is why it is virtually impossible to pick up TV signals from distant TV stations. If TV broadcasts in the United States are to be received in Europe, for example, they must be beamed via a relay satellite in orbit. Therefore, planetary colonists in exploratory missions would have to work through such satellites or via the ISTs. Whether the use of TV would be justified in the circumstances is debatable. If geologists came across rock outcrops or formations on the planet's surface indicative of the existence of rich mineral deposits they might wish to relay pictures of these back to base for scrutiny by experienced mineralogists.

Electronic forms of communication, of course, require power sources. If for any reason power were not available the colonists would have to retreat to a rudimentary and pre-twentieth century level of communication. If our own fossil fuels here on Earth were exhausted we would be compelled to fall back on nuclear energy. In a century or so, however, it is not unlikely that our descendants will have devised satisfactory methods of harnessing the inexhaustible energy of the Sun and of the ocean tides. There are many practical difficulties to overcome but, given time and the necessary will, both sources of energy probably will become exploitable. Colonists should have the necessary knowledge and expertise by then to harness the tides or their alien planet or—which would be even more attractive—to utilize the light energy emanating from their central star or "new Sun."

8

The Search for Minerals

THE KIND AND QUANTITY of mineral deposits found on their new planet will be a question of absorbing interest to the colonists. If a civilization is to have a modern technology, minerals—especially metals—must be found in abundance. There also must be the ability to extract them and then smelt or otherwise remove the metals from their natural ores. Without metals indigenous to the planet, the colonists would have to create an agrarian civilization and eventually, for lack of anything better, would find themselves using wooden plows. Even after locating mineral deposits some time must inevitably elapse before they can be mined in quantity, refined from their ores, and transformed into usable metals. For a period the colonists might be forced back to some of the techniques and standards of an earlier era, the possibilities and results of which are discussed in a subsequent chapter. Mineral deposits are therefore of paramount importance.

Minerals represent the "building blocks" of rocks. Some rocks are made up of only a single mineral such as quartz, whereas others contain many. Concentrations of minerals containing economic quantities of useful metals such as tin, copper, tungsten, iron, lead, and chromium are known as ores. They are found in a variety of guises. The various ores of the same metal can have entirely different origins. Magma (molten rock beneath the surface) is the origin of many mineral deposits. Some are formed within the cooled, consolidated igneous mass itself. Due to magmatic segregation the minerals become highly concentrated, and this results in the formation of a rich ore deposit. On Earth there are many examples of this, including the rich iron deposits in northern Sweden, chromite deposits in South Africa, and nickel (as sulphide) in Ontario, Canada. During cooling of the magma, hot gases and liquids sometimes are forced under considerable pressure into the surrounding rocks. As these highly mineral-rich solutions cool and pressure is reduced, the minerals are deposited as solids. On other occasions the hot liquids are forced into small cracks in the surrounding rocks where, on cooling, they produce veins, such as those of calcite (calcium carbonate). If there is a collection of veins (generally referred to as a lode) it may contain economically important minerals as well as worthless ones, known as gangue. Though veins have no great thickness they often run for considerable distances and can penetrate to great depths.

Igneous rock pushing its way up from a magmatic source also changes the surrounding rocks because of the heat. This is especially true in instances in which direct contact has occurred. In cases where hot, gaseous, mineral-rich solutions replace some of the original rock, rich deposits of copper, zinc, and lead ores are formed. All of these metals would be very useful to an expanding terrestrial colony. Interactions of this nature tend to occur at points where granite and limestone have come into contact, granite, being an igneous mass, providing the heat. On the other hand, ores that result from hot, aqueous solutions are known as hydrothermal deposits.

At times mineral-rich solutions act preferentially, replacing only certain elements in the original structure. At other times the whole mass may be affected and changed. This is largely due to the specific crystalline solutions of the intruding metal and in what respects they compare with those of the element being replaced. A description of this process would be too involved for our purposes, but numerous works on geology and mineralogy discuss the subject in depth.

Replacement of this nature often leads to the formation of rich and extensive deposits. A classic example is the rich deposits of pyrites at Rio Tinto in Spain and in the "copper belt" of Zambia. Pyrites (chalcopyrites) is a combined sulphide of both iron and copper. It looks like gold and for that reason is popularly known as "fool's gold." Its crystalline structure is quite different from that of gold. Pyrites looks very brassy compared to gold.

The type of mineral deposit is determined by the temperatures of the solution and of the associated gases. Moving away from the zone of greatest heat, the order of the minerals likely to be deposited follows the sequence tin, copper, lead, zinc, and iron. This does not mean that all of these will necessarily be present. Nevertheless, colonists finding copper might expect also to find deposits of tin and lead on either side of it. Hydrothermal solutions at times surface in the form of hot springs depositing the metals they carry in solution. A typical example is the mineral cinnabar, a compound of mercury and sulphur, from which the liquid metallic element mercury is obtained. Cinnabar is regarded by geologists as a "low temperature" mineral, but this description is purely relative. The spray from one of these hot springs seems hot enough if it impinges on any part of the human anatomy!

It should not be thought that hot, intruding igneous masses such as granite or gabbro are the only forms of mineral deposits. Rocks on the surface of the Earth are subjected to continual weathering, especially by rain. As the water percolates downward it may leach (dissolve) useful elements from the upper rock layers through which it passes. These are borne downward to the water table and there deposited. Several rich deposits of copper have been created in this way, the rain passing through low-grade upper deposits of the ore and carrying and then concentrating copper ions at a lower level. Wet, tropical conditions also have an effect on silicate rocks containing aluminum, breaking them down and in the process producing deposits of bauxite, the principal ore of aluminum. Some iron ores as well as deposits of manganese are produced in this manner.

When rock suffers degradation due to weathering, the resulting particles are washed into streams and rivers. Within these particles there can exist a variety of minerals. Some are not easily changed and generally being of a heavy variety drop to the bed of the stream or river as placer deposits. When the word "placer" is mentioned, most of us automatically think of gold. It is hard to say what value colonists from Earth will place on this precious yellow metal. It has many technological uses but most of it is melted and run into molds to become ingots of pure gold. The metal's rarity as well as its attractiveness ensure its use for monetary purposes. Having taken it out of the ground at great expense, we generally put it in bank vaults and bury it again. This sometimes seems rather silly. The bulk of Earth's tin comes from placer deposits in Malaya and Indonesia.

Many other sedimentary rocks (rocks laid down by the action of water) are important sources of minerals, including gypsum, common salt (sodium chloride), and potash. Gypsum has been produced largely by the evaporation of ancient, wholly enclosed seas; thus it is termed an evaporite. As these ancient seas evaporated under the hot rays of the Sun, the least soluble minerals were the first to be precipitated. Potash salts, being the most soluble, were the last to come out of solution. In this manner the great halide deposits of Stassfurt in Germany were formed.

All this has, of necessity, been a very brief résumé of how mineral deposits are formed. But now two fundamental questions have to be answered:

1. What guarantee is there that the same essential minerals would be found on an alien planet, no matter how Earth-like that planet happened to be?
2. Precisely what kind of features would geologists look for with respect to particular minerals?

The answer to the first question is that the planet, being Earth-like, would almost certainly have been formed from a primordial nebula in the same manner as Earth and from the same elements. Like Earth, we would expect it to comprise an inner core, an outer core, a mantle, and a relatively thin crust. The inner core of Earth is believed to be solid and to consist of a nickel-iron alloy. The outer core is probably fluid and may be composed of iron and iron sulphide. The mantle is certainly a mixture of iron and magnesium silicates. How did these divisions form? This is of considerable interest, because a planet of similar type to Earth at a similar distance from an identical type of star is virtually certain to have a like interior. There are two theories, one or other of which is almost certainly valid. One is the theory of homogeneous accretion, the other the theory of heterogeneous accretion.

The first of these is as follows. The Earth was formed when elements in that portion of the primordial solar nebula where the

Earth now is condensed into small, solid particles. These drew together as a consequence of gravitational attraction to form a protoplanet. At the start this protoplanet was relatively cold and homogeneous. The particles that had come together to form it were made up of a number of elements, but each particle had more or less the same constitution. The proto-Earth was a mass of particles; though each in itself heterogeneous, they had joined together to form a protoplanet. Thus, though the elements in the protoplanet were a mixture, this mixture was the same throughout the protoplanet. In this respect the protoplanet was homogeneous, being composed of a mass of the same heterogeneous particles—almost a contradiction in terms!

Melting now started to take place within the proto-Earth, and since all the particles forming it comprised materials of differing density, layering started. The heavy elements sank toward the center—hence the formation of dense iron-rich cores and mantles of less dense materials. This mode of planetary formation calls for a powerful heat source to promote the melting process (1,000 to 1,500°C). Possible sources have been attributed to gravity and to radioactive materials. Impacting materials, as they built up the planet, would have their kinetic energy partly converted into heat. This heat would be proportional to the strength of the gravitational field and would increase as more and more fast-moving particles of condensed matter slammed into the growing protoplanet. Radioactive heating depends on the concentration of heat-producing radioisotopes in the embryonic planet, but calculations indicate that this would not be as great as the heat due to gravitational impacts.

The alternative to homogeneous accretion is heterogeneous accretion. The essential premise of this theory is that iron-rich cores accumulate first and only later become surrounded by the layer of silicate material that we call the mantle. The implication here is that the primordial solar nebula was so intensely hot that all the elements were in the gaseous state. As its temperature fell, iron condensed before the silicates. Hot, heterogeneous planetary accretion provides a means of producing layered structures directly from the primordial solar nebula, thus removing the immediate need for internal heat sources. It is, however, unlikely that the Earth's original heat gained in this manner would be adequate to maintain its dynamic convecting mantle and its fluid outer core. We are still compelled to invoke as additional heat sources gravitational and/or radioactive heating. On balance geologists and astronomers tend to favor the homogeneous theory rather than the heterogeneous one. The structure of Earth-like planets of stars in other parts of the galaxy is in all probability almost identical. Such planets were formed from other primordial stellar nebulae and this means that their geology should be very similar indeed.

Now to the second question. Given the above circumstances, what types of planetary feature would geologists relate to possible mineral deposits? First, let us look at some of the arrangements of minerals on our planet. Certain elements in the crust of the Earth (and therefore in Earth-like planets) are enriched in the crust because of their geochemical properties. These are termed lithophile elements (from the Greek *lithos* meaning stone). They occur in the "stony" layer of the Earth that we know as the crust. They are all elements with a fairly strong affinity for oxygen, occurring as oxides or silicates of potassium, sodium, aluminum, calcium, strontium, rubidium, zirconium, and barium. These are all enriched within the crust, or outer layer of the planet. Magnesium, though also classified as lithophile, is nevertheless more abundant in the mantle than in the crust.

Some elements have a strong affinity for the element sulphur and because of this form sulphide ores. These are known as chalcophile elements. Copper sulphide is a classic example (hence the name, which is derived from the Greek *khalcos,* meaning copper). Many chalcophile elements occur in conjunction with copper sulphides, including zinc, lead, cadmium, and silver.

A third group, the siderophile elements (from the Greek *sideros,* meaning iron) normally occur in metallic form rather than in ores chemically bonded with oxygen or sulphur. These include nickel, palladium, platinum, rubidium, and gold. These elements are concentrated in the core of Earth-type planets either in their central, metallic region or in their outer, sulphide-rich region. It is hardly surprising, therefore, that most siderophile and chalcophile elements are depleted in the crust. It looks as if the colonists from Earth are not likely to be stumbling over great chunks of gold or platinum.

The reasons why the elements have these varied affinities is because of their electron/nucleus configurations. This must be taken on trust by the reader but if he or she would like to know more about this, any decent textbook on elementary geochemistry will supply the information.

However, wc still have not answered how geologists would know where to look for deposits of valuable and essential minerals. The answer is contained in a single phrase—by geological prospecting (though, as we will shortly see, more sophisticated techniques are not only possible but also highly advantageous). In seeking mineral deposits there is always the possibility of finding an ore body at the surface, but this entails a lot of luck and should never be relied upon. What is much more likely are indications at the surface sufficiently promising to warrant drilling operations. The sorts of geological evidence that would justify this can be best summed up as follows:

1. The widespread presence and extent of common rock types with which certain types of mineral deposit are known to occur. This does not mean they are certain to occur. Evidence of this sort must therefore be supported by more definite indications of mineralization obtained by other means.

2. The presence of specific rock types known to have originated by internal processes that frequently lead to mineralization. This is more definite than the preceding point of evidence but still does not constitute a definite guarantee of mineralization. Nevertheless, geological evidence of this kind could be sufficiently encouraging to warrant further exploration and employment of techniques to be described shortly.

3. The discovery of surface concentrations of economically important minerals, or (as is much more likely) the products of surface alterations of such mineralization.

   Take, for example, the surface products beneath which lies a rich deposit of copper sulphide. If the region happens to be humid and damp it would be perfectly reasonable to anticipate the oxidation products of sulphide minerals. The most likely would be hydrated ferric oxides. Where oxidation is not complete, copper sulphides may be visible. The green and blue oxidation products of copper sulphide deposits (generally known as "blooms") are extremely distinctive and, if in sufficiently high concentration, can be seen very easily from a distance during an aerial survey of the region. Other minerals showing distinctive blooms are cobalt (pink), vanadium (orange), and uranium (green, yellow, or orange). The last-mentioned would surely be of very special interest to colonists. Moreover, it is by no means uncommon for certain elements to occur in association with one another (for example, lead and zinc with copper, nickel with cobalt, vanadium with uranium). Thus the discovery of one of these so-called blooms might indicate the presence of more than one element in economic concentrations. Moreover iron-rich blooms are oxidation products of the iron sulphides pyrite (FeS_2) and pyrrhotite (FeS), both of which are common associates of many other sulphide (and some oxide) ores.

The above are, on the whole, relatively unsophisticated techniques about which there is a strong element of hit or miss. They are hardly the methods 21st-century colonizers and explorers from Earth would use if they had the physical resources to use more sophisticated methods. Indeed, more sophisticated methods exist and are widely used today. With the passage of another century or so they will be further refined. A great deal of involved equipment is not generally required, and it will be of interest to have a brief look at these techniques.

MAGNETIC METHODS

Superimposed on the global and regional pattern of Earth's (and, we would suppose, of an Earth-like planet's) magnetic field are local magnetic anomalies produced by variations in the intensity of magnetization of rock formations. Part of the magnetism in rocks is induced by the present magnetic field of the Earth, while part is inherited from past magnetic fields generally referred to as remanent magnetization dating from the time of formation of magnetic minerals in the rocks. The polarity of magnetization in a mass of rock is thus cumulative due to remanent magnetization plus induced magnetism due to Earth's present magnetic field. As a consequence, the magnetic polarity of the rock mass will not necessarily have the same orientation as the present magnetic field of the Earth.

This brings us to the factors governing the intensity of magnetization in rocks. One of these is the magnetic susceptibility of a rock, and the other is the magnitude of the Earth's magnetic field. The magnetic susceptibilities of rocks—that is, the extent to which they are affected by magnetization—are determined by the amount of magnetically affected mineral within them. Four of the most common magnetically affected minerals are magnetite, ilmenite, pyrrhotite, and haematite. The element common to all four, as might be expected, is iron (all are ore oxides of iron except pyrrhotite, which is a sulphide of iron). Ilmenite also contains the valuable element titanium. Common silicate minerals possess extremely low magnetic susceptibilities. Consequently, most sedimentary (water-laid) rocks have extremely low magnetic susceptibilities; igneous and metamorphic rocks, medium magnetic susceptibilities; while basic and ultrabasic rocks (low-silica SiO_2 content) have, not surprisingly, very high magnetic susceptibilities. Thus any rock containing magnetic minerals such as those

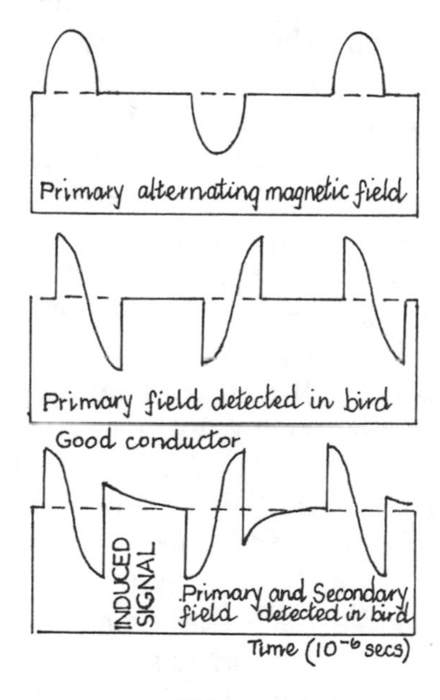

FIG. 3.

Transmitted and received electromagnetic wave forms over a conducting ore mass. The secondary electromagnetic field effectively distorts the signal received.

mentioned above produces a magnetic anomaly, and an instrument designed to measure magnetization will show an unduly high reading over a rich iron ore deposit. However, this is all very well in the case of Earth, where magnetism induced by its magnetic field and that due to remanent magnetization are known. Geologists on another planet would first have to deduce these two factors. This would take time and an almost total exploration of the planet. Therefore, the method might not commend itself to these geologists, though exceptionally high magnetic anomaly readings still would stand out against the planet's natural background magnetism and would therefore be indicative of the proximity of an ore deposit.

ELECTROMAGNETIC METHODS

Though electric currents are able to flow through the Earth, nearly all the common rocks are extremely poor conductors of electricity. Two factors are capable of enhancing the flow of electrical current within them. One is the presence of ionized ground water, which effectively reduces the resistivity of the rock. This in itself is unlikely to indicate the presence of metallic ores. The second is the presence of metallic sulphides, which possess good electrical conducting properties. Since metallic ore bodies are natural conductors of electricity, sufficiently strong electromagnetic radiation produced externally will cause these ore bodies to act like natural antennas, causing alternating electric currents of the same frequency to flow within them. This is analogous to radio waves from a transmitter setting up electric currents of the same frequency within a receiving antenna, as discussed in Chapter 7. Electric currents also produce magnetic fields and so secondary alternating magnetic fields are induced in metallic ore bodies. In practice the procedure is to have a primary alternating field produced by coils in an aircraft through which alternating current is fed, while the secondary magnetic field due to an ore body is sought by another set of coils trailed behind it like a towed glider (generally termed the "trailing bird"). As the detector coil passes through the induced magnetic field of an ore body an alternating current is generated within it. For terrestrial colonists on a remote Earth-like planet such a technique would surely be of considerable value in their search for essential minerals.

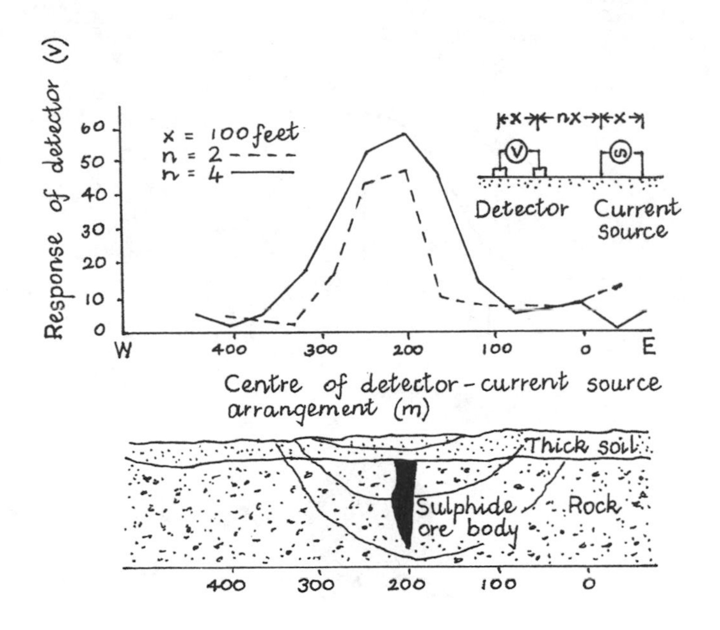

FIG. 4.

Induced polarization profile over vertical sulphide ore mass. Inset shows the arrangment of current source and detector.

ELECTRICAL SURVEYS

It has already been stated that rocks are very poor conductors unless they happen to contain ionized ground water or, less commonly, metallic sulphides. Because even wet rocks have higher resistivities than many metal ores, metalliferous mineral deposits can produce what are known as electrical anomalies. If two electrodes are pushed into the ground and a voltage applied, current will flow between them and, provided the ground is not dry, this current is capable of penetrating quite deeply. However, should an ore body be present, current, because of much-reduced resistivity, will flow more easily and the presence of the ore body will be detected. In practice the technique is slightly more involved than this, but it illustrates the basic essentials. It is certainly a technique that geologists on a terrestrial type planet could usefully employ.

It has been possible in these pages to present only a rough general guide to the techniques used to locate hidden mineral

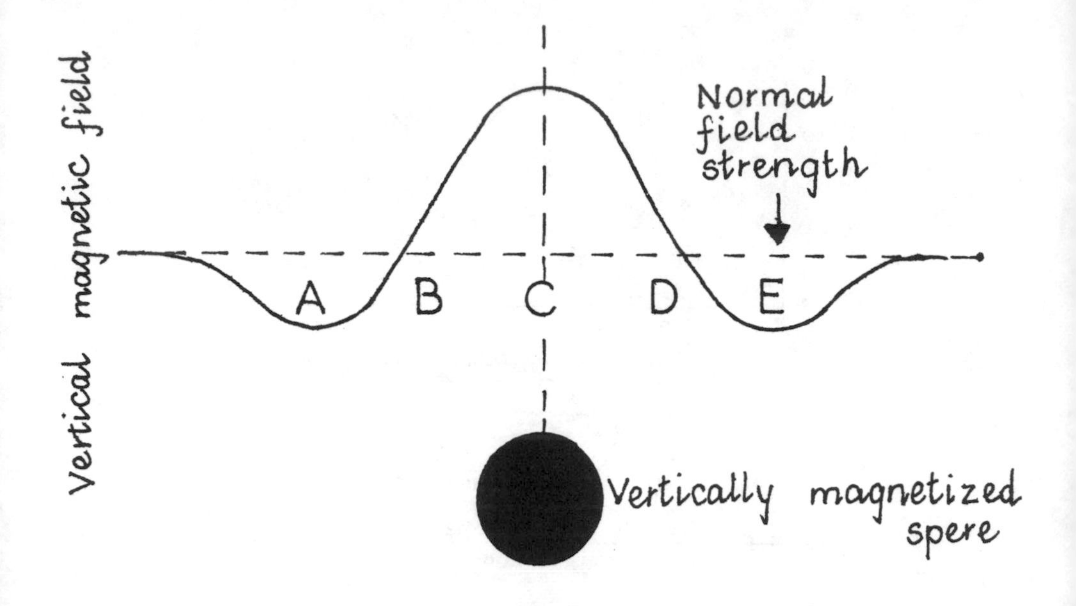

FIG. 5.

Magnetic anomaly of the vertical component over a magnetized sphere.

deposits. We have assumed that the planet is, to all intents and purposes, akin to Earth, as only a planet of this nature would be appropriate to extensive colonization without the colonists having to live under impossibly artificial conditions. And since vast stocks of minerals cannot be carried in the ISTs, it is essential that minerals and metals be found in abundance. Of course, finding them is one thing, and extracting them from the ground and refining them for use is quite another. Nevertheless, all this was achieved on Earth, and on "Nova Terra" it surely could be achieved again. For decades colonists from Earth might have to improvise, and they would probably be compelled to revert to some of the techniques of bygone ages. However, Rome wasn't built in a day and neither will a highly technological civilization on a new world light-years distant.

Extraction and processing of minerals more or less epitomizes the unique problems facing planetary colonists in their early years on the planet. For that reason the next chapter is devoted to this aspect.

9

Mineral Processing

THE PREVIOUS CHAPTER provided the briefest of insights into the techniques colonists on a world akin to Earth might employ to locate mineral deposits. But finding essential minerals is just the tip of the iceberg. No matter how extensive the deposits are, they are totally useless as long as they remain in the ground. To be of value they must be extracted and then processed. At this juncture the problems start.

It is difficult to predict the expertise and the hardware that our descendants a century or two from now will possess. Nevertheless, although it is a civilization able to send its representatives to other stars (albeit with difficulty), it is not going to be able to transport much of the results of its superior technology in the form of bulky industrial equipment. What it will be able to take along will eventually wear out. It will not be possible to replace this until a sound technological and industrial base has been created, and this may take time. Because of these factors colonists with the knowledge of their century may be compelled to employ the methods of late-20th-century Earth or even of earlier periods. It is rather as though we asked an airline using supersonic jets to revert for an indefinite period to piston-engined, propeller-driven machines—or even to gliders and hot air balloons. In science fiction it is all so very easy—a great fleet of mammoth ISTs reach the planet and unload everything that could possibly be required and that in vast quantities. In no time at all great cities of the future and vast industrial complexes arise. That would be very nice, but it is impossible to reconcile with the real situation an early colonizing expedition from Earth will face.

A close look at the requirements of modern mineral extraction and processing illustrates what the colonists will be up against in their early years on the new planet. They are going to be faced with a kind of "chicken and egg" situation. A planetwide technology cannot be created until metals and minerals are freely available, yet until a planetwide technology exists, securing and processing the minerals is possible only on a rudimentary scale. Of course, the next few centuries may see machines whose capacities we cannot yet dream of. Back around 1936 an excellent film version of H. G. Wells' *Things to Come* was produced. In that film, after mankind had at last seen the error of its ways and renounced war forever, it produced a wonderful new technological civilization. Among the many marvels to which viewers were treated was the spectacle of great machines that could level hills and gouge great chunks of rock and mineral out of the ground. This was science fiction, of course, and no doubt all the real difficulties and problems were glossed over, but nevertheless one was left with the feeling that all this might one day come to pass. However, the real problem is that there is a limit to the number of great machines ISTs can carry—and that brings us back to square one.

MINING: OPEN-PIT METHOD

An open-pit mine sounds simple if we regard it as a big hole in the ground, but as is so often the case, there is a great deal more to it than meets the eye. This would prove as true on an Earth-like planet as on Earth itself. It is not simply a matter of going to the region under which the mineral lies and commencing to dig a great hole. It is essential that the shape of the hole be tailored to suit both the geology of the ore body and the limitations of the mining equipment used. The colonists may very well have brought more sophisticated equipment with them than is presently available to us, but it will be strictly limited in quantity and will not be replaceable for a long time after it wears out—and it is well established that mining operations are very rough on equipment.

During the life of a mine (the period during which there remains sufficient ore to justify its existence) the original hole must be progressively deepened and widened. This has to be taken into account during the design stage. Planning the evolution of a mine is an exceedingly complex operation, and fresh problems are likely to be encountered, rendering constant revision and modification necessary. Therefore, in spite of the urgency of the colonists' requirements (and they would be quite urgent!), haste will have to be made slowly.

Most open-pit mines on Earth have a series of steplike terraces running around the entire excavation. Typical examples of this may be found in copper-ore mines in Arizona. In all open-pit mines the terraces are carefully planned, since the angle of slope of the sides of the pit is controlled by the properties of the rock in which excavation is made. In pits excavated in strong rocks the slopes can safely be made more steep than in weak rocks. In weak rocks these slopes must be made very gentle to eliminate the possibilities of collapse. It then becomes necessary to remove a much larger amount of waste rock to reach the actual ore body. This absorbs much time and effort and calls for a lot of equipment and a lot of people.

With the pit designed, the colonists would only then begin to dig, an operation that can be conveniently subdivided into three well-defined stages. The first of these is drilling and blasting, which consists of drilling holes, inserting conventional explosive charges, and firing them. If, however, it is a bauxite (aluminum ore) mine, blasting and drilling are necessary only rarely, for this is normally a very soft and easily worked material. Aluminum is a very useful, strong, lightweight metal, and obtaining this ore could present much less of a problem to colonists with severely restricted facilities. Unfortunately, separation of aluminum from its ores requires a relatively prodigious amount of electrical power, and it seems likely that this will always be the case.

After blasting, the fragments of ore are collected. At present on Earth this is done by very large electric shovels that must, nevertheless, be maneuverable in view of the terraced nature of an open pit. Finally the ore must be transported to the processing and smelting plant. At present huge diesel-electric trucks are used. Colonists from Earth, if they have the equipment, can clearly get their ore. Most of this equipment could come with them from Earth. But could the vitally necessary processing equipment? We will address ourselves to this later.

UNDERGROUND MINING

This is a highly skilled operation that has a mystique and vocabulary of its own. It is hard to say in what direction it will develop in the centuries to come.

An underground mine is a vastly more complex and dangerous entity than an open pit. All that follows would also apply to colonists on another "Earth," and the factors that might defeat them in their efforts should be fairly obvious. The layout of an underground mine is dictated by the shape of the ore body. The planners must devise a three-dimensional framework of workings that will afford access to as much of the ore body as possible. They must also be extremely careful to ensure that the actual operation of mining will not cause the workings to collapse. Despite care such collapses still do occur, with catastrophic loss of life. Drilling and blasting underground are much more difficult than on the surface, not just because of the difficulties of working in a confined space but also because sufficient time must be allowed after the firing of each charge for fumes to clear before the broken ore can be removed. In addition, explosive charges used underground have to be relatively small, with the result that not a great deal of ore is produced by each blast (the amount is estimated at a few tens of tons). This ore then has to be transported from the blast site to the shaft and then finally raised to the surface and removed to the processing plant, which ought to be located as close to the mine as possible. It seems clear that this would present a tremendous challenge to recently arrived planetary colonists, but it also is well to remember that here on Earth some mine workings date from an early period (for example, certain copper mines in the vicinity of Lake Superior, Canada, seem to have been intelligently and systematically worked long before the arrival of white people). This mine could not in any way be compared with a modern copper mine, but it does show that a primitive people were able to engage in the deep mining of a mineral ore. Colonists to another star system would certainly not be primitive, and with such instruments of their technology as they were able to transport from Earth they could perhaps achieve more in this respect than we imagine, but it still would be a considerable challenge to attempt underground mining in the early days of colonization—especially if the necessary equipment had been worn out in other operations.

PROCESSING

The stages involved vary widely from one ore to another, so here we can only generalize. Different kinds and grades of ore require different treatment—a fundamental truism in mineral processing. The three main processing operations are crushing, separation and concentration, and smelting.

Crushing appears a reasonably straightforward process involving the mechanical breakdown of large chunks of ore into small fragments. In practice it is not quite that simple. Since the main objective of the entire processing operation is to free the valuable particles of ore from the rock surrounding them, it is vitally important that the crushed material have the right degree of fineness. If it is insufficiently fine, then subsequent concentrating processes will not prove effective; if it is crushed too finely, much of the ore will be rejected along with rock powder. The type of equipment used for crushing ore is fairly simple and would certainly not occupy too much space in the great ISTs during the long journey from Earth.

Separation and concentration can be achieved in a number of ways, none of which are particularly involved or necessitate vast technical resources. The principal techniques used are magnetic sorting, gravity separation, and flotation separation. These are unlikely to worry colonists too much. When we come to smelting, however, the picture is very different, for this process requires items such as large furnaces, feeders, air-blowers, and much other ancillary equipment. Refining involves the use of yet more furnaces. If the metal concerned were copper and this was intended for electrical purposes, it would need to be of very high purity, which would necessitate the employment of an electrolytic copper-refining unit.

It would appear, therefore, that colonists are unlikely to be in a position to extract and process substantial amounts of metallic minerals for quite some time unless new and as yet undreamed-of processes have by then been discovered and perfected. This does not seem likely, and several generations of colonists may have lived and died before their technology has attained a level equal to that which the original colonists left behind on Earth. That would appear to be the only realistic assessment. Only rapid-transit techniques such as were outlined in one of this book's predecessors, *Interstellar Travel: Past, Present and Future,* would permit the rapid development of contemporary, terrestrial technology on the new world. It is not possible to say if such techniques will ever be feasible. An orthodox colonizing mission would only have the equipment and essential supplies it could carry. For the most part colonists from Earth would have to be prepared to "live off the land" in the colony's formative years. Life would not be easy for these people at first; for pioneers it never is. They would have left the sophisticated technology of Earth for the wild, untamed wilderness of a new Earth. In many respects they would have, for a number of years, a simpler life. That might prove a blessing. Already life here on Earth is becoming increasingly complex and the increasing incidence of nervous and mental disorders may be a consequence of that.

10

Food and Agriculture

IT COULD BE ARGUED QUITE convincingly that by including chapters dealing with minerals and metals prior to one on food supply we have been guilty of putting the cart before the horse. Colonists from Earth to an Earth-like planet of another solar system must eat. True, considerable quantities of food could have been carried in the ISTs and, since it has been possible to preserve and greatly prolong human life by cryogenic processes, the same presumably could be done for foodstuffs intended for early consumption and for seed that could (hopefully) be sown on appropriate parts of the planet's surface. But no matter how much food was carried in the great ISTs, sooner or later (and in the circumstances probably sooner) it would be consumed. A very high degree of priority would need to be given to indigenously produced food, failing which the colonists would soon starve. Livestock brought from Earth (assuming they survived the journey) would require time to breed. Slaughtering them right away would be tantamount to eating the corn seed.

It is perfectly conceivable that some of the planet's indigenous flora and fauna might be fit for human consumption, but this is not a premise that can be safely relied on. Those bright red berries that tasted so delicious could be harmful, or fatally poisonous, to our kind. Moreover, flora containing many safely edible species might not constitute the balanced diet essential to the health and well-being of human beings.

Since it is our firm belief that only a replica or near replica of

Earth could possibly serve as a permanent, full-scale terrestrial colony, we must assume that the soil would in many if not all respects be similar to our own. All forms of life on land depend directly or indirectly on the soil. Soil itself is the cumulative result of physical and chemical weathering of the underlying barren rock and can vary in depth from a few centimeters to several meters.

Biologists would begin by examining and analyzing the alien soil and determining whether it could nourish and sustain terrestrial food crops. If so, then the sooner ground was cleared and crops sown, the better. Crops provide not only food for humans and livestock; crops also produce the seeds that will provide fresh crops—the food of future years.

Soil, as has already been stated, is the result of time and weather upon rock, but strictly speaking this does not yield soil as such but only a mass of unconsolidated, inorganic particles. This becomes soil in the accepted sense only when it acquires a minimum organic content in which plants are able to take root and deposit their litter. This organic content in the case of an Earth-like planet would hopefully have been supplied by its indigenous plants. These would then provide the necessary nutrients for the growth of terrestrial plants brought from Earth.

As organic matter accumulates, fine humus builds up in upper soil horizons. If we consider the profile (section) of a typical soil sample we find that it reaches down from the most recently deposited topsoil to the parent bedrock, revealing various easily recognizable layers or "horizons." Beneath a layer of gray-black topsoil and its upper layer of humus there lies a layer known as subsoil that, though poorer in organic materials, is richer in accumulated minerals than the topsoil and humus above it. These minerals are important to the growth of terrestrial food crops, and biologists among the colonists would be anxious to ascertain whether the subsoil of the planet contained these in sufficient concentration. The horizon below this is made up of the partially weathered particles of the lifeless parent bedrock.

The type of climate dictates the type of soil and this, in turn, determines the crops that will grow best. The biologists will make visits to both temperate and tropical regions of the planet to secure soil sample sections for examination and analysis. An acid brown soil is typical of temperate climates. In the case of wet, cool climates there is likely to be a layer of fresh litter and humus followed by a leached acid horizon overlying a mineral humus layer of enrichment. A thick, red soil containing iron compounds is common to humid, tropical regions where chemical and biological activity are both high. These examples are typical of different climatic regions on Earth, and on a terrestrial-type distant planet there would be some differences, but as long as the planet had a similar axial tilt and a not too dissimilar length of day and seasons, the soil would be very similar to that of Earth.

Soil formation is the result of the interaction of parent rock, land relief, time, climate, and decay. The parent rock represents the source of most soil. Weathering physically reduces some of it to a mass of gravel, sand, silt, and clay. It must be emphasized, however, that soil is not always identical to its parent material because of the numerous chemical transformations and physical disturbances it is likely to experience during its formation. Rainwater falls upon it, seismic activity rearranges it, and earthworms and bacteria play a part.

Another factor influencing the creation of soils is land relief. On steep slopes only thin, dry layers of soil accumulate because of rapid water runoff, while in level, high country soil, layers containing clay tend to accumulate. In poorly drained regions where organic decay is slow, thick layers of dark, organic soil builds up. Again, these are factors likely to prove common on all the other millions of "Earths" scattered throughout the galaxy.

Still other factors are to be considered. A hillside receiving direct sunlight will have soil different from that which is hidden from direct exposure because of the inevitable differences in moisture content. Another passive influence in the formation of soils is time. Young soils have hardly had time to acquire distinct horizons. Older and more mature soils will have acquired a well-defined profile—one that undergoes only minimal changes with the further passage of time.

Probably the single most important factor in the development of soil is climate. Water in the form of rain is essential to all chemical and biological changes in soil. As it slowly percolates it leaches the surface layers, depositing material in the subsoil. Thus in regions of extremely heavy rainfall such as the tropics, a high degree of upper surface leaching takes place and the soil is consequently rendered relatively sterile. In hot, arid climates excessive evaporation takes place, resulting in high concentrations of silt deposits in the rock. As might be expected, the rate of chemical and biological activity in soil is directly affected by temperature. Such activity is high in tropical regions. Decay is consequently rapid and the soils are poor in humus. In tundra regions, the topsoil is likely to be frozen for more than six months in a year, while the subsoil remains permanently below the freezing point. Under such conditions organic matter accumulates in thick layers. Lateritic soils provide an excellent example of the effects of climate. In hot, wet environments such soils become highly leached and contain little other than deposits of bauxite (aluminum oxide) and iron. If directly exposed to fierce, tropical sunlight these soils are transformed into a hard-baked mass known as laterite to which vegetation can never return.

Earth therefore produces many types of soil—some useful, some utterly useless as far as crop growing is concerned. The main parameter, as we have just seen, is climate, and this to a large extent is dictated by Earth's axial tilt of 23½ degrees from the plane of the ecliptic. This tilt is largely responsible for the existence of tropical, temperate, tundra, and arctic conditions. It is highly improbable that colonists from Earth would find a planet with precisely the same amount of axial tilt. If the difference were merely a few degrees and the length of day roughly similar, the colonists probably would find zones of similar extent, and the type of crops growable in each particular zone probably would correspond roughly with those in the analogous zones of Earth. But if that tilt were appreciably greater—say, 40 degrees or so—summers in the planet's northern and southern hemispheres would be hotter than those on Earth, and winters would be very much colder. Grain crops would not therefore be located as they are on Earth. If the axial tilt were even greater the planet might be Earth-like geologically but certainly not climatically. It could still be regarded by terrestrials as habitable and were it the only Earth-like planet in the solar system to which it belonged, it would probably have to be accepted. Consider an analogous but hypothetical case. Members of an alien civilization from a sun like our own and a planet lying at roughly the same distance from it reach the Solar System on a strictly one-way colonizing mission. They soon find that Venus is intolerably hot and the atmosphere unbreathable. The planet from which they come is about the same size and mass as Earth. Its atmosphere also is very similar. Unfortunately, their native planet has an axial tilt of 50 degrees, which renders its climate and seasons very different from Earth's. It is not wholly suitable, but since they cannot go back whence they came and there is nowhere else to go, it must suffice.

There is a further aspect in which the soil of the chosen planet must be akin to that of Earth. Our soil represents a very complex ecosystem; a square meter of terrestrial soil teems with more than 1,000 million individual forms of life, ranging from microscopic organisms to insects and worms. All play a highly important part in helping to aerate the soil and to accelerate the vital processes of decay and humus formation. Without these lowly creatures (which many regard with repugnance) there would be no soil for agricultural use or at best only a very low-grade type. This would have a most adverse effect on the growing of crops and the production of food. A planet fit for human occupation must consequently be one the soil of which contains a similar concentration of life serving the same purpose as does its equivalent on Earth. Alien "creepie-crawlies" would no doubt differ somewhat in external characteristics from those of Earth, but on an essentially Earth-type planet on which evolution had followed a roughly similar course it seems reasonable to expect creatures performing a parallel function. If the planet is to be able to grow terrestrial crops or their alien equivalent, lowly creatures like these are going to be necessary. According to one recent estimate the humble earthworm alone is capable of turning over between one and ten tons of soil per 100 acres each year. This is certainly no mean feat for such a creature. By eating and excreting the soil they also effectively change its texture and composition.

The role of soil bacteria is also crucial, for not only do they bring about the fixation of nitrogen from the air in a form usable by plants, they also promote the processes of decay.

Under normal conditions here on Earth soils replenish themselves naturally. At times, however, retrograde agricultural practices prevail and have the effect of reducing the fertility of the soil. Colonists will be aware from the start that they should avoid such practices. Probably the most disastrous of them lead to a "dust bowl." In a "dust bowl" the essential topsoil has been entirely removed by the wind during prolonged periods of drought; this is mainly due to the felling of trees in a region that requires their protection. In the early 1930s several parts of the United States fell victim to this.

The colonists would be intent from the outset on maintaining the soil at the highest possible productive level. This requires a combination of all the techniques of soil science in the preparation of the land; irrigation and fertilization will be carefully attended to, and the right crops must be chosen. The choice of crops is essential to ensure the stabilization of the soil and the prevention of erosion. The biologists will have to work hard and fast, and no doubt laboratory facilities will be provided for them in one of the ISTs.

In an earlier chapter we dwelled briefly on the water requirements of Base Camp I, but now it is necessary to consider the subject in the wider context of the irrigation of food crops. Clearly these will not flourish if water is lacking. Presumably, as on Earth, there would be temperate regions where irrigation was totally unnecessary because of adequate rainfall. There would also very likely be regions where crops could flourish successfully if only water were able to reach their roots in adequate quantity. Some might wish to argue that surely by the 21st or 22nd centuries food crops might be superfluous if the old science-fiction concept of daily bodily requirements being met by a strictly synthetic diet has become reality. This concept still seems strangely unreal, but even if it became feasible the supply of "food pills" would eventually be exhausted unless facilities for manufacturing fresh stocks had been established—and this would bring up the problem of obtaining the essential chemicals. For our purposes, we shall assume that food still will be necessary and chemical nutrition not be a viable proposition even at that future date.

Irrigation is the artificial watering of land and may be necessary for the following reasons: rainfall is too sparse to produce crops; rainfall is either seasonal or too erratic; a natural supply of water must be augmented to increase crop yields.

On Earth, irrigation has been responsible for the transformation of vast expanses of arid, infertile land into highly productive soil. The desert has *literally* been made to bloom! At the present time some 400 million acres of land are being irrigated.

Food would be required by colonists soon, since limited amounts brought in a preserved, refrigerated state from Earth would be consumed relatively quickly. The region around Base Camp I would be the first to be cleared and have crops sown. As the number of colonists increased, the need for food would grow. As colonization began to spread across the planet with the number of mouths to feed steadily multiplying, more and more crop-growing areas would become essential. Some of these areas would almost certainly be barren due to lack of water. As on Earth, irrigation would be the answer if these areas were potentially arable, and it is a reasonable assumption that so long as plant and equipment were available, techniques used on Earth at the time of the mission's departure would be those used by the colonists on their terrestrial-type planet. These, in their essentials, would probably be somewhat similar to those in use at the present time and fall into five distinct categories:

Underground irrigation. This brings water to the plants directly through the soil. This method is only practicable on those parts of the planet where the terrain is level and the soil both highly permeable and overlying an impermeable layer that traps ground water, thus permitting it to seep upward to the plant roots by capillary action. This method of irrigation has the advantage of minimizing water loss due to evaporation, though it does tend to deposit unwanted mineral salts at the surface. Accumulations of this kind would have to be cleared from the soil, and this is achieved either by heavy rains or by deliberate flooding. To be of use to colonists this form of irrigation would be practiced only where the following conditions are met:

1. level terrain
2. highly permeable soil
3. a lower impermeable layer
4. high rainfall
5. if rainfall is not high, there must be a large body of water nearby that can be tapped, thus creating deliberate floods of moderate duration.

Surface irrigation. This is much the more common method, the procedure being to dig long furrows in the land to be irrigated by directing water into them from a single source. Water flows out over the land in a broad sheet, and the surplus is drained off at another point. This method tends to be most favored for growing fodder crops. Colonial communities engaged in the rearing of beef cattle probably would adopt it.

Basin irrigation. For colonists whose staple diet happens to be rice, this method would be highly appropriate. Water is trapped by low retaining walls around the edges of the growing area until the soil becomes saturated.

Furrow irrigation. This is probably the most suitable method when growing grain. Grain is the generic name for a number of crops including maize, wheat, rye, and barley. This seems a likely technique on another "Earth" 4 or 5 light-years distant. Furrows about 1,600 feet long are plowed between the rows of grain. These slope gently away from the water source so that water running along them does not cause excessive erosion but is enabled to soak gradually into the soil around the crop. The major disadvantage of this method lies in the difficulty of ensuring that all parts of the growing area receive an equal amount of water. In ensuring that all parts receive an adequate supply some receive an excess resulting in a waste of valuable water.

Overhead irrigation. This simulates natural rainfall by the use of spray lines or sprinklers. These can be so adjusted that the artificial rainfall can be made to vary from a fine spray mist to a near tropical downpour. The sprinklers generally are arranged in rows and connected by pipes to a central water source or pumping unit. The main advantages claimed for this method are:

1. the land requires no special preparation (noteworthy from the viewpoint of planetary colonists) and
2. the flow of water can be controlled. This is quite important, since too much water can be as damaging to food crops as too little. This method on a simplified and greatly reduced scale is used in many parts of the world to water lawns during a period of drought.

Another method colonists could use on their new planetary home would be "dry farming." This permits the farmer to grow his crops without irrigation in regions where the annual rainfall would normally be too low for successful cultivation. (On Earth this is regarded as below 20 inches.) The technique is absurdly simple and proves as effective, under the right conditions, as complicated and sophisticated techniques of irrigation. Olive trees, or other trees of a similar type, are planted at the center of shallow basins, which funnel rainwater inward and downward. Would this work on another planet? There seems no reason why it would not as long as the topography (the shallow basins) resembled those on Earth.

The colonists might want to try irrigation on a grand scale. A question that appears in so many aspects of colonizing a planet light-years from Earth is: Would the colonists have the facilities? Grandiose schemes of irrigation might have to wait for years or even decades, until the colonists had created the necessary technological and industrial base. The know-how needed to create large irrigation schemes and great dams is one thing, the material capacity to do so is another. The sort of project we are considering would be akin to the Imperial Valley in the United States, where a whole area was totally transformed by large-scale irrigation engineering. At one time the region was completely arid. Now it is intensely farmed. The colonists could hardly afford to concentrate their limited machinery and resources on projects of this size, and in their first few years on their new Earth they would be well advised to steer clear of regions requiring *extensive* irrigation.

We must now consider another facet of the food problem. On Earth pests and diseases are estimated to destroy nearly one third of the world's crop harvest per year. Crops are ravaged at various stages in their growth and during reaping and storage by insects and other pests, by diseases, and by weeds and useless grasses that compete with crop plants for vital nutrients. Mammals and birds also take their toll. A third of the world's crops destroyed or rendered useless every year represents a very high proportion, and this despite the widespread use of pesticides and herbicides. It seems extremely probable that colonists raising crops on their new planetary home would also find a vast range of indigenous pests and diseases that might prove immune to terrestrial pesticides and herbicides. Even if they did not, stocks of these materials would not be inexhaustible, and it would take some time for biologists and bacteriologists to identify the myriad pests and diseases and probably even longer to test and develop suitable pesticides and herbicides or set about manufacturing the old ones. Many of these chemical compounds have elaborate molecules and their manufacture calls for considerable facilities. As before, mere expertise is not enough. However, before this century people survived for many thousands of years without pesticides and herbicides. It is very likely that the first colonists on another Earth would have to do the same.

Surely it would be sensible to consider the indigenous flora (and perhaps fauna, too) as sources of food. Edible crops peculiar to that particular planet might be more appropriate than our transplanted terrestrial crops. This ia a possibility that surely will not be overlooked.

Once again, however, time enters the equation. We are envisaging a virgin world (hopefully devoid of intelligent forms of life) that is in a totally uncultivated state. The nature of its flora and fauna is completely unknown. To identify and catalog its plant and animal (including fish and bird) life is going to take time. Ultimately certain flora could easily prove edible and this, of course, applies to some of the fauna, too. In all probability, however, years of research will be required before this opportunity is fully exploited.

Land clearance for the sowing of terrestrial or indigenous crops would call for a maximum effort. It is obvious that only a small number of earth-moving machines could be brought in the ISTs. These may break down before the industrial base to manufacture more is complete. In that case a great deal of arduous human labor is going to be required. It would be simple to revert to the "slash and burn" system, once largely employed on Earth but now virtually superseded. Centuries ago it was used extensively in northern Europe and now is still used in parts of central Africa. In the northern European example, first all trees were cleared from the ground earmarked for cultivation. After the cut trees were somehow removed (presumably the roots being for the most part simply left in the ground) the brush was cut and the entire area burned to remove all further vegetation. Fairly high yields (by the standards of the times) of rye and oats were eventually harvested. After a year or two the soil had been so denuded of its natural minerals that the land was allowed to revert to scrub and the nomadic people moved on. For the tribes of that day and age the system, though inherently bad, had certain merits. Planetary colonists with nomadic inclinations might also find it useful.

What of hydroponics? Couldn't this be used as an interim measure? At the present time hydroponics does not seem to have made great advances, though its future potential, within certain strictly defined limits, seems bright. It is essentially a technique for growing plants over water instead of in soil, suitable chemicals being dissolved in the water that feed the plants through the dipping roots. It has been successful in growing plants in totally soilless conditions. In hydroponic agriculture all a plant's needs including oxygen, light, water, mineral salts, and other nutrients are artificially provided within the protected and temperature-controlled environment of a greenhouse or similar structure. These plants are freed from competition with weeds, and damage due to insect pests and viruses is greatly reduced. The input in hydroponics comprises carbon dioxide, oxygen, water, nutrients, energy, light, and heat while the output comprises potash and alcohol from burned waste. The alcohol can subsequently be burned to produce energy.

It is difficult to envisage hydroponics being developed on a large scale as it is highly artificial and, on the whole, quite expensive to set up. Here on Earth it still makes more sense to utilize the countless acres of arable land, and on an Earth-like planet this would be true as well, though much virgin forest and grassland would first need to be cleared. Hydroponics are well suited for the growing on a small scale of such items as tomatoes. It is extremely difficult to envisage the technique ever vying with the growing of wheat on the great prairies of the United States and Canada. Where food is required in relatively small quantities by a few people, the technique has something to commend it. It might flourish in a space station where all conditions are artificial anyway, or perhaps someday on the Moon or Mars. Our colonists light-years from Earth might, however, find it a useful adjunct to those food sources available in the early days of the colony.

The matter of clothing should probably be mentioned here. Extensive stocks will have been carried, and these will no doubt prove adequate for a time, being depleted less rapidly than food. At this later point in history it is extremely likely that most if not all clothing will be manufactured from synthetic fibers. Such garments are, at the present time, not always without their disadvantages, but perhaps over the next century or so the disadvantages will have been overcome. However, we must wonder whether a sufficient industrial base will be established in the first few years of colonial life to permit the manufacture of synthetic fibers. There is no certainty that it will, and some of the colonists could well be forced to adopt raiment of animal or vegetable origins such as sheepskin coats, leather jerkins, or woolen stockings. It may seem odd that people able to bridge the gap between stars, coming down by shuttle craft from great ISTs in fixed orbit, and wearing the futuristic garments of a century or so hence may eventually, if only for a relatively short period, be forced to don apparel more reminiscent of the sixteenth and seventeenth centuries. This could be just one of the ironies and idiosyncracies of the peculiar situation in which they find themselves. In a later chapter we will dwell at length on the "back to the past" theme.

11

Ocean and Atmosphere

IN DESCRIBING THE HYPOTHETICAL planet reached by the equally hypothetical colonizing mission from Earth we have been taking certain small liberties. A planet may be very Earth-like, but it can hardly be an exact replica of our own in every respect. We can visualize a planet of more or less the same dimensions, gravity, atmosphere, and temperature range that orbits a solar-type star at roughly the same distance as that separating Earth and Sun. The elements it contains within its crust, mantle, and core are akin to those of Earth. Further than that we really cannot go except to add that its geological history may have been very similar to that of Earth. If the first colonists from our world reach such a planet they will regard themselves as singularly fortunate although, as was outlined in an earlier chapter, a planet of this nature must have been their hope from the outset. Colonizing a world so distant from Earth is going to be no easy matter, but the more it is like Earth, the easier it will be.

One of the main reasons why an Earth-like planet cannot be a total facsimile of Earth is because of the oceans that cover so large a proportion of our own world. This is not meant to suggest that an Earth-like planet would not have its quota of oceans and seas, too. It would. (It could not really be classed as Earth-like if it did not.) But there would almost certainly be differences. The oceans and seas of Earth constitute nearly 75 percent of our planet's surface. In the case of an Earth-like planet, this figure could be significantly less—or more. Moreover, the positions of these seas and oceans in relation to those of contiguous landmasses would inevitably be very different. Of still greater importance would be the matter of tides.

The tides on Earth, a feature to which we are so accustomed, are caused by the gravitational influence of Sun and Moon; but the greater influence is that of the Moon for, though so much smaller than the Sun, it is nevertheless much nearer. Suppose, however, that this Earth-like planet had either no large satellite or, like Mars, only a pair of very minute ones. The only appreciable tidal effects in either case would be due to the star around which the planet orbits. It is also possible that the planet could have a significantly larger satellite than the Moon, or more than one large satellite, in which cases tidal effects would be appreciably greater.

Clearly we cannot present oceanographic details of a purely hypothetical planet. However, we can explore the relationship between the ocean and the climate here on Earth and try to relate this to an Earth-like planet.

Oceans and atmosphere are inextricably linked, and this can be attributed to the continuous transfer of momentum, energy, and matter at the ocean-atmosphere interface. Oceans absorb a high proportion of the solar radiation penetrating the terrestrial atmosphere, thus acting as a heat reservoir. As a consequence they heat up slowly in summer and cool down slowly in winter. Thus they render the climate more equable by storing some of the summer's heat and releasing it during winter. The atmosphere receives not only heat but also most of its water content from the oceans. Therefore, climatic changes are closely related to oceanic changes.

The circulation of oceanic water involves both surface currents and circulation at depth. The latter is due to cold, dense water sinking in the polar regions and then moving toward the equator. One extremely important aspect of oceanic circulation is that it results in areas of dense fish population along some continental coasts. Water wells up from deeper levels to replace surface waters transported away from the land by surface currents. This upwelling water is extremely rich in nutrients due to decay of organic matter sinking from the surface. These regions therefore have a high biological production, and though on Earth they constitute a mere 0.1 percent of total oceanic area, they account for 25 percent of our world's annual fish-catch. In Chapter 10, dealing with food and agriculture, all the emphasis was on land-based crops, but here there is a rich and immediately available store of valuable food. As on Earth some varieties of fish might be more acceptable (or palatable) than others, but it would still be reasonable to expect a fairly wide variety of edible species. Of course, evolution is an involved process, and to expect an exact parallel with what happened here on our home world would be expecting a lot. There would probably be a number of divergences.

However, all this has little or no relevance to the interaction of oceans and atmosphere. The question of tides and their magnitudes constitutes as good a starting point as any. Tides on Earth result, in the main, from the differing gravitational attractions that the Moon and the Sun exert on the surface of our planet. It is at times hard to believe, as we stand on a beach or rocky shore and watch the tide slowly ebb or flow, that bodies a quarter of a million miles and 91 million miles distant, respectively, are responsible.

The initial factor with which it is necessary to come to terms is known as the equilibrium tide. The gravitational attraction of the Moon just balances the centrifugal force resulting from the rotation of the Earth about the center of mass of the Earth-Moon system. Because of the large difference in mass between Earth and Moon, the center of mass about which the system revolves is located within the Earth itself (see Figure 6). Though this centrifugal force is the same for a unit mass at any point on the Earth's surface, the gravitational attraction varies according to the distance of the unit mass from the Moon. Consequently there is a net force directed toward the Moon at the point on the Earth's surface closest to the Moon as well as a net force directed away from the Moon at the point on the Earth's surface farthest from the Moon. At other positions on the Earth's surface there are smaller net forces directed at some angle to the vertical (see Figure 7). If the ocean surface were to reach equilibrium with these forces it would have to become perpendicular everywhere to the resultant of this force and the Earth's gravitational attraction. This equilibrium tide would form an ellipsoid "bulging"

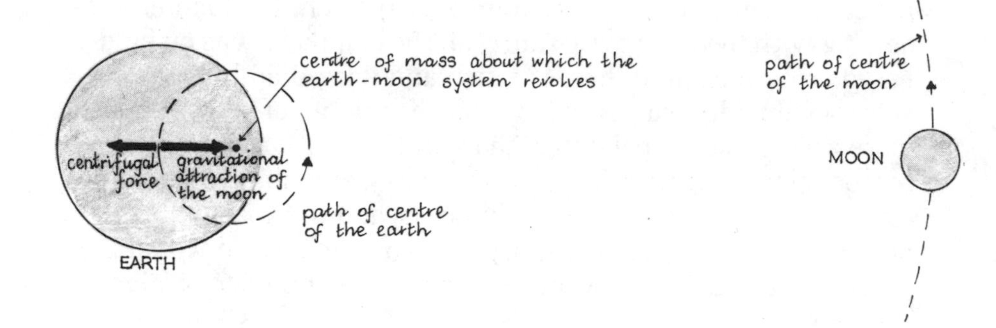

The balance of the gravitational attraction and centrifugal force for the Earth-Moon system.

FIG. 6.

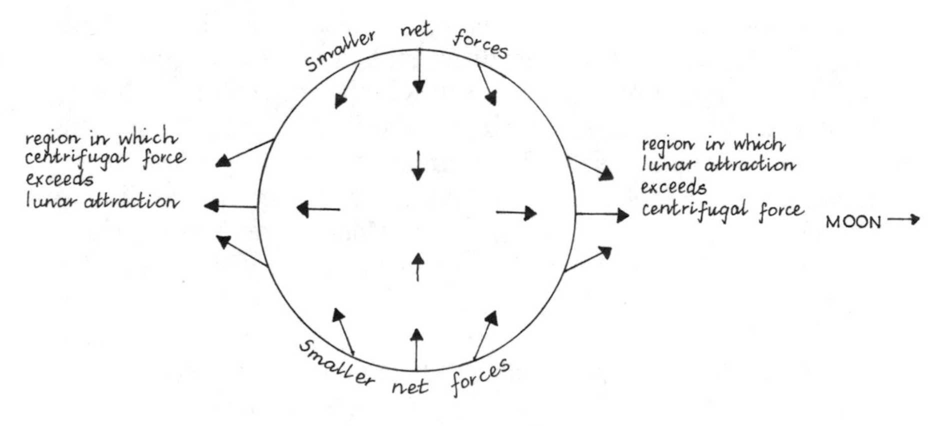

The pattern of the main tide-producing forces over the Earth's surface.

FIG. 7.

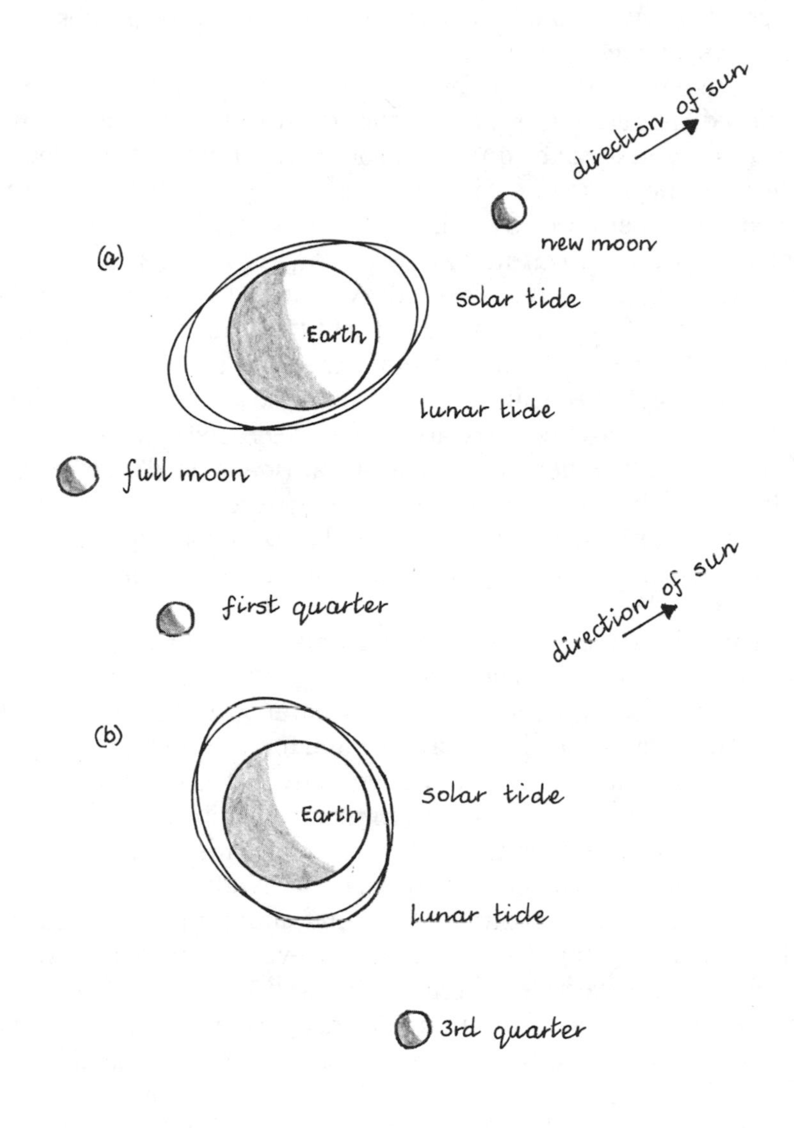

FIG. 8.

Diagramatic representation of the solar tide superimposed on the lunar tide to produce (a) spring tides, and (b) neap tides.

where the gravitational attraction of the Moon was greatest and least, respectively.

Because of the rotation of the Earth this situation is never achieved. To retain the shape of the equilibrium tide the oceanic "bulges" would have to travel around the Earth at a velocity whereby they remained in the same position relative to the Moon—in a period of 24 hours, 50 minutes, the period of the Earth's rotation about its axis with respect to the Moon. Even were the Earth entirely covered by water, this would still be possible only if the depth of water permitted the "bulges" to keep up with the Moon's rotation around the Earth.

The concept of equilibrium tides does, however, explain their periodicity. In essence there are two wave crests every 24 hours, 50 minutes. Consquently the basic lunar tidal period is semidiurnal (every 12 hours, 25 minutes). These two crests are not of equal magnitude unless the Moon is directly above the equator. The declination of the Moon (see Figure 8) varies during the month by as much as 28½ degrees.

As previously mentioned, the Sun also exerts tide-producing forces at the Earth's surface. Were it not much more distant than the Moon, it would exert a much greater gravitational effect. Since the magnitude of the gravitational forces is directly proportional to the mass of the attracting body but inversely proportional to the cube of the distance, the solar equilibrium tide is only 46 percent as great as that of the Moon. This produces tides having periods of 12 and 24 hours, respectively. When these are superimposed on the lunar constituents, maximum tides (generally known as spring tides) are produced every 14 days at new and at full Moon, respectively, because Sun and Moon are then acting in unison. Conversely, minimum (or neap) tides result when the two bodies act in opposition to one another. Since the declination of the Sun changes and because there are also other regular variations in the relative positions of Sun, Earth, and Moon, including the distances separating them, the tides show a sequence of variations. All these must be taken into account by compilers of tables showing times of high tide at various points on the Earth's surface. This was never the simplest task, but computers have now made the computations easier.

Therefore, if an Earth-like planet has no large moon relatively close to it as does Earth or only perhaps two tiny ones such as Mars possesses, then the tidal pattern of that planet's oceans and seas is going to be entirely different from that to which we are accustomed. Tides will be due almost entirely to the star around which the planet orbits. Our Sun contributes only 46 percent of the tidal effect, so if star and planet are comparable to Sun and Earth in the distance between them, then the planetary tides would be much less. There is also the opposite possibility. The new planet might have a bigger moon than we possess and therefore much greater tides or two or three medium-sized moons, leading to a highly involved tidal pattern. Earth-like though another planet may be, the tidal pattern is certainly going to prove rather different. How would this affect the colonists from Earth? That question is best answered after we examine the effect of tides on Earth.

The main effect of tide-producing forces is the creation of waves in oceans, seas, and bays with periods that coincide with those of the major tidal constituents. In the majority of places the semidiurnal period is the dominant one, but in some parts the diurnal period predominates. Tide-generating forces comprise two components, one vertical and the other horizontal; the horizontal component is mainly responsible for the creation of waves. In some instances these behave as progressive waves that travel across ocean basins at velocities determined by the depth of the particular basin concerned. When these eventually reach coastlines, and especially when they enter marginal seas and bays, they become subject to reflection, and a counterwave in the opposite direction is created. If these counterwaves are in phase with the waves causing them, tidal oscillations of considerable amplitude result—or in simpler terms, very big waves are generated. A notable example of this process is provided by the Bay of Fundy in eastern Canada, where the waves often get higher than 40 feet during the spring tides. The analysis of waves and their patterns is a complex matter and not appropriate here, but the effect of waves on coastlines can be considerable and, given the right combination of time and circumstances, quite disastrous. In brief, the Moon and the Sun produce tides that in turn produce waves, and these affect shorelines. Colonists from Earth on reaching their new planet should view all coastlines with circumspection, and until a detailed study has been made it would be highly advisable not to locate base camps or settlements too close to them.

The type of tides discussed in this chapter have nothing to do with the "tidal waves" mentioned in Chapter 5. The term "tidal wave" is a popular misnomer. Such water movement generally results from submarine earthquakes or great eruptions of island volcanoes such as the one that occurred on Krakatoa in 1883. The normally accepted term for such geologically produced waves is the Japanese word *tsunami*. Normal tide and wave patterns can produce catastrophic effects. This was highlighted spectacularly in January 1953 when an extensive area of eastern England suffered immense devastation. This coastline is washed by the waters of the North Sea. Progressive tidal waves impinge upon it from the turbulent Atlantic Ocean. Some of these come down from the north, and others pass through the more restricted Straits of Dover in the English Channel; this renders the tidal pattern both complex and potentially dangerous. Similar land and sea configurations are perfectly possible on other planets.

The River Thames flows out via a broad estuary into the region where the waters of the English Channel merge with those of the North Sea. In recent years the prospect of a phenomenally high tide surging up the Thames and inundating large parts of London has been causing the authorities serious concern. The stage seems to have been set for a disaster of mammoth proportions, and consequently many millions of pounds have recently been spent on the construction of a sophisticated barrage by which such a surge could be blocked. At the time of writing, this has just been completed. A barrage of this nature raises immense technical problems, not the least being the fact that the Thames is one of the greatest waterways in the world, and it is essential that the passage of shipping not be impeded. The situation is compounded by the fact that most of southeastern England is gradually sinking. Terrestrial colonists and their descendants must keep such problems in mind when they construct their future cities and ports lest they court disaster. Much will depend on the magnitude and inherent nature of tidal forces and therefore on the size and distance from the planet of any satellite or system of satellites. In science fiction a familiar theme has been that of Earth's destruction by collision or near collision with an intruder planet or asteroid entering the Solar System from interstellar space—an unlikely event at best. In the early 1930s two American authors, Philip Wylie and Edwin Balmer, wrote an enthralling and blood-chilling novel, *When Worlds Collide,* on this theme, and in the 1950s this was produced as a highly spectacular movie bearing the same title. The film illustrated vividly the colossal tides created in the oceans of Earth by the close passage of a large planetary body. The streets of lower Manhattan began to fill with water, and the flooding continued until only the tops of skyscrapers emerged from the waves. Colonists from Earth are unlikely to suffer such a disaster, but they might discover that their planet was orbited by a very large satellite. In that case, not only would high tides render coastal areas of the planet most inhospitable, but also gravitational strains could easily lead to an undesirable level of seismic and volcanic activity.

It is more likely that Earth-like planets orbiting other stars will have only small satellites or none at all. Of the rocky planets in the Solar System, Mercury has no moon. Neither does Venus, while Mars has only two tiny moons. Pluto is now known to have a satellite, but whether this planet is rocky or just a great sphere of frozen gases remains an open question. Of the rocky planets, Earth is unique in having a large satellite only a quarter-million miles distant. Many astronomers now regard the Earth and the Moon as a twin planet system rather than as a planet and a satellite. Possibly such a relationship is rare, but how rare it is we cannot say. Only when the age of interstellar exploration dawns will colonists discover whether their new planet has one satellite, or more than one, or none at all.

Our Sun (and others of the same stellar type) emit energy, the three main components of which are:

1. Infrared radiation, accounting for 48 percent of the total
2. Radiation visible to the human eye, accounting for 43 percent of the total
3. Ultraviolet radiation and X rays, accounting for 9 percent of the total

Were our planet devoid of atmosphere, the amount of radiation received by its surface per day would depend only on the length of time it was exposed to the Sun's rays, the angle between the Sun's rays and the surface of the Earth, and the distance separating Earth and Sun. These factors vary both with latitude and season and the Earth is some 2 million miles closer to the Sun during the depths of the Northern Hemisphere winter. Earth does possess an atmosphere, however, and because of this the type and proportion of the Sun's radiation received at the surface of our planet varies somewhat from that outlined in the foregoing list. This is due to absorption and scattering as it passes through the various layers of atmosphere. The shortest-radiation wavelengths (highest frequencies) are absorbed by gases in the upper atmosphere, a feature that results in certain important photochemical reactions. This absorption is extremely fortuitous, for overexposure to ultraviolet and X rays is highly inimical to the health of the human body. On a hot, cloudless summer day a fair amount of ultraviolet radiation reaches the Earth's surface, and any reader who has basked for too long in such sunlight wearing only a swimsuit will soon be aware that he or she has been unwise. Not only will there be the discomfort of sunburn, but feelings of sickness and general malaise can also be present. (The day need not be cloudless, especially in the tropics, for the latter effects to be felt if one is hatless.) Ultraviolet radiation is not to be trifled with.

By absorption of ultraviolet and X rays, molecules and atoms of upper-atmosphere gases lose electrons, thereby becoming positively charged ions. That layer of the atmosphere in which the concentration of ions and electrons is greatest lies 60 to 300 kilometers above the Earth's surface, is known as the ionosphere, and, as pointed out in Chapter 7, permits the long-distance transmission and reception of radio waves by reflecting these back to the surface of the Earth.

A further manifestation of the effect of ultraviolet and X rays on our planet's upper atmosphere is the transformation of oxygen into ozone. Ozone is formed mainly above 40 kilometers, though it is also found in considerable concentration between 20 and 35 kilometers. This gas completes the absorption of ultraviolet radiation. The transformation of oxygen to ozone is brought about by the dissociation of oxygen (O_2) into two single atoms ($O + O$). These single atoms combine with unaffected oxygen to produce ozone.

$$O + O_2 = O_3 \text{ (ozone)}$$

It would be to the advantage of colonists on a terrestrial-type world for this reaction to be proceeding in its upper atmosphere.

In Earth's lower atmosphere the only gaseous constituent capable of absorbing significant amounts of the solar radiation that still remains is water vapor. Some 10 percent of total solar radiation is absorbed by it. We can see how the existence of seas and oceans, resulting in the presence of water vapor in the atmosphere, helps control the amount of radiation received on the surface of a planet from its sun.

Absorption of radiation is not the only process taking place. Reflection also occurs in the atmosphere when solar radiation is incident on clouds. The proportion of radiation reflected by clouds is termed the albedo, and the magnitude of this factor is dependent both on the type of cloud and on its thickness. On average, so far as Earth is concerned, this is about 55 percent. Here is yet another example of the indirect effect of seas and oceans on a planet. Calm water itself provides a very low albedo (only about 2 percent). It is the clouds, caused by evaporation and subsequent condensation due to water, that produce the real reflective power.

The amount of solar radiation reaching the Earth's surface may be as much as 80 percent of that impinging on the top layers of the atmosphere when the sky is clear or less than 20 percent when it is overcast. Variations are also caused by the amount of water vapor and dust in the atmosphere and by the length of the radiation's path through the atmosphere, the latter being a direct function of the Sun's elevation in the sky. Any large scale volcanic eruptions can significantly influence the amount of dust in the atmosphere and therefore the amount of solar radiation reaching Earth's surface. In this respect therefore any large scale volcanicity can greatly influence Earth's climate. This was made manifest by the tremendous eruption of Mt. St. Helens in the United States in May 1980. Will Earth-like planets out among the stars experience similar volcanic outbursts? There seems no reason why they should not if their geological histories approximate those of Earth.

It must be obvious that if Earth (and planets like it throughout our galaxy) continued to absorb solar (stellar) radiation without any attendant loss of heat, their surface temperatures would steadily rise to an alarming degree. This does not occur because these planets emit electromagnetic radiation into space. There is thus a balance between incoming solar radiation and outgoing terrestrial radiation. If, however, annual mean values of incoming and outgoing radiation at a specific location on the Earth are determined, it is virtually certain that an imbalance will be found. This is because processes other than radiation lead to the transfer of heat, particularly in the atmosphere and ocean.

Remaining incoming radiation is absorbed at the surface of our planet and, by analogy, we may assume this applies also to other Earth-like planets. This absorption is due either to land or to water and brings us to the second major point of this chapter: heat transfer between atmosphere and ocean.

In addition to the Sun's heating of the atmosphere by the emission of long-wave radiation, the atmosphere also receives heat by conduction across the interface of either land and air or water and air. This is followed, as might be expected, by convection within the atmosphere. Normally conduction would lead to very little transfer of heat, but in the circumstances the process is required only within a layer above the ground and air or water and air of a few millimeters or less. Vertical motion of the air resulting from thermal convection then transfers the heat upward, thus maintaining an adequate temperature gradient at the interface for conduction to continue fairly rapidly.

Where a water surface—be it an ocean, sea, or large lake—is involved, another important heat-transfer process takes place. This has the effect of transferring heat from the Sun into the atmosphere. Though couched here in basically physical terms it is one with which most of us, unless we reside in desert or near-desert regions, are all too familiar. Heat from the Sun impinging on a water surface causes evaporation, and this is followed by condensation in the atmosphere. For each gram of water that evaporates from the surface of an ocean or sea, a specific amount of heat is required (for the mathematically minded this is 2.47 x 10^2 joules). Water vapor is carried into Earth's atmosphere as latent heat, and this is released when condensation takes place.

Heat is lost from the oceans in the following proportions:

by long-wave radiation to atmosphere and space	41%
by conduction and condensation	5%
by evaporation	54%

Evaporation is therefore paramount, with conduction and condensation the least. To produce this pattern the average ocean surface temperature must be greater than that of the overlying air. Of even greater import, the vapor pressure of the air overlying the ocean must be less than the saturated vapor pressure of air at the temperature of the surface water so that evaporation takes place. Occasionally the positions are changed. For example, near the Grand Banks, just off the coast of Newfoundland, air temperature in the spring exceeds sea-surface temperature. This results in the transfer of heat from atmosphere to ocean so that condensation occurs at the ocean surface and just above it. The result is fog, often very dense, and prior to the coming of reliable radar this rendered the region most perilous for shipping. Because it is receiving heat from the atmosphere the water at the surface becomes warmer and therefore less dense. The overlying air, on the other hand, becomes cooler and denser.

Another process known as advection or horizontal transfer of heat is necessary for a net loss of heat by radiation in high latitudes and for a net gain by radiation in low latitudes. Were this not to occur, temperatures in equatorial regions would increase by about 10 percent. In polar regions they would decrease by more than 20 percent. The implications would be far-reaching. The proportion of our planet's surface covered by ice and snow would increase considerably. In turn this would increase the albedo or reflecting power in middle and high latitudes, thus bringing about even more cooling in these regions. It is estimated that about 80 percent of this heat advection is occurring in the atmosphere where the global winds carry warm air and water vapor with its latent heat toward the poles. It has also been estimated that the contribution of the oceans averages 40 percent of the total poleward.

As with so many facets of our theme, the subject of the oceans and their effects on a terrestrial-type planet could be continued almost indefinitely. In this chapter we have sought to focus attention on the most salient points to illustrate, within a limited space, the relationships between ocean and climate as well as those between tides and close-lying celestial bodies. Because of these features another "Earth" could not possibly be a complete facsimile of our own. For this to occur the geography would need to be identical as well as the planet's relationship to its central star and satellite, which would themselves need to be identical to our Sun and Moon, respectively. Such a planet may exist somewhere among the distant star systems, but it is against all the laws of chance that our colonists would happen upon it. It is said that if an infinite number of monkeys were let loose on an infinite number of typewriters sooner or later one, unknowingly, would produce the complete works of William Shakespeare. The odds against finding a planet exactly similar to Earth are roughly the same.

12

Law and Government

SO FAR WE HAVE DWELT almost exclusively with the technical and material aspects of colonizing an Earth-like planet light-years from Earth. It is time to consider the government, laws, and social structure that such a colony would have in its early years.

A number of years ago I wrote *Journey to Alpha Centauri,* which featured a 200-year journey to a hypothetical planet of Alpha Centauri, using the concept of "generation" travel. I was criticized by one reviewer for envisaging a totalitarian type of society aboard the great ISTs *Columbus* and *Drake.* On serious reflection I felt obliged to adhere to what I had written. The reviewer stated that, like so many scientists, the author "could not conceive of a pioneer society in space that was not totalitarian in control of mating, reproduction and attitudes." While fully respecting the reviewer's opinion, I failed to see how such a mission could succeed without a strict measure of control over the lives of the eight generations involved. I was envisaging a "closed" society totally divorced from Earth and heading for an unknown world. If each person had been allowed to go his or her own way the result would have been the total failure of the mission.

Pioneers from our world to another "Earth" 4 or 5 light-years distant would certainly exist under different circumstances from those cooped up for a lifetime within a giant star ship. Nevertheless, they would still face tough odds. A reasonably tight rein would need to be kept on the lives of each of the colonists by some form of governing body. Without such control the mission could founder, and the lives of the colonists be lost.

It might in the circumstances be a good idea to ponder the term "totalitarian." The dictionary defines it as "relating to a one-party dictatorial form of government under which the freedom of the individual is nonexistent, every aspect of the life of the nation being state-controlled." This defines "totalitarian" in its nastiest form and fully applies to Nazi Germany and to several countries of the world at the present time, the most notable example being the Soviet Union. The idea that the state was made for man and not man for the state is heresy in these lands.

The form of rule that would exist in a traveling space colony aboard a great star ship would have to be quite authoritarian but at the same time it could be humane and benevolent. This is not a contradiction in terms as long as the ruling body ruled in the best long-term interests of the people over which it was set. During World War II, both the United States and Britain found it necessary to restrict and control the lives of their citizens. It was authoritarian rule but, in the circumstances, absolutely necessary. Moreover, it was seen to be necessary by the public and therefore freely accepted. Franklin Roosevelt and Winston Churchill exercised considerable power but were still at all times answerable to their peoples. The same could not be said for Adolf Hitler or Joseph Stalin.

The governing body of a terrestrial planetary colony, in view of all the difficulties and perils, would need to have a more decisive power than we normally allow our democratic governments today. Authoritarian is probably a better and more appropriate term than totalitarian to describe its style. As time passed, as technology developed, as numbers increased, and as colonists (or more likely their descendants) spread out ever farther over the planet's surface, relaxations could come about until at last, hopefully, democratic rule along the lines of that in the United States, Canada, and the countries of Western Europe could be established. What would need to be avoided is a government that had become attached to the principles of fascist or communist ideology. Over the past six decades, millions have sacrificed their lives to the exorbitant demands of those two doctrines. One of them still holds sway over many lands, purporting to champion egalitarianism while in fact doing nothing of the sort.

Egalitarianism is all very well in theory, but it makes no allowance for human nature. All men may be born equal but there is no reason they should remain that way. I do not mean to suggest that grinding poverty should be allowed to exist side by side with vast wealth, privilege, and ostentatious luxury. This can only lead to class strife and bloody revolutions in which the downtrodden of yesterday become the tyrants of tomorrow. We must attempt to keep the mistakes and evils of Earth from being transferred to another world. In social orders, as in all facets of human activity, perfection will never be achieved, but it is nonetheless desirable to strive after it.

When we examine the history of mankind we are confronted by a lamentable and appalling catalog of evil, cruelty, greed, and intolerance. As civilization has advanced through the centuries these characteristics have not altered. Technologically mankind has made great strides. Would that he had made the same strides morally. The two are badly out of phase. During the darkest days of World War II the late Winston Churchill, in one of his famous speeches, spoke of "the perils of perverted science." If our race continues to advance technologically but to remain more or less static in other respects, then science could become increasingly perverted, the servant of corrupt men, enabling them to control the thoughts and activities of whole populations. George Orwell in his novel *1984* described the sort of society we must do our best to avoid.

We cannot expect human nature to change overnight—even on a new and unspoiled planet. In many respects it may never change. But it is to be hoped that mankind, reaching for the stars, will leave his worst evils and excesses behind. In man there is also much good, much humanity, and much compassion. Much thought must therefore be given to the type of society that will be set up on a new world far from Earth. It seems desirable that it should not see, as Earth has, the formation of separate nations continually at each other's throats, squabbles among differing faiths, race fighting race, and society divided sharply and calamitously between rich and poor. This may seem to be that demand for perfection mentioned a few lines back. It may not be wholly achieved, but it should be striven for.

What then should be the style and aims of the governing body of a new, developing colony on another world far removed from Earth? Clearly a policy of laissez-faire dare not be adopted. Reasonably strict discipline would at first have to be exercised over the colonists, though we must hope it could avoid becoming tyrannical. If these young men and women had quit Earth to escape the tyranny of some particular political "ism," there will be no point in reestablishing something similar when they arrive. The new colonists will want to establish a free and just society even if the freedom is curtailed a bit at first. Perhaps if we think back to the early years of the colonists in the New World on Earth we may find the basis of an answer.

The colonization of other planets to escape political persecution is paralleled by the experiences of those men and women who two and three centuries ago determined to free themselves from religious persecution. Harried in Europe by the infamous Spanish Inquisition and by a body in England known as the Court of High Commission, these people were glad to venture into the unknown. Unfortunately, the colonies they set up in the New World had no prospect of trade with or succor from Europe. They had to survive in subsistence conditions to begin with. Over the course of two centuries the Spanish, British, French, Portuguese, and Dutch established colonies in the Americas. Many of the original settlers were seeking freedom, but, as time passed, they were followed by the representatives of the governments they had left behind, and, in typical European fashion, the mother countries were soon at each other's throats to see who was to dominate the Americas. In North America the British came out as the winners in this sordid business but then made the fatal mistake of trying to compel the colonies to submit to taxation without representation. The colonists deeply resented this, and the result was the Declaration of Independence in 1776, the American Revolution, and the birth of the United States of America.

There are interesting parallels in this to what may happen when we colonize the stars. Unlike the early settlers in America, the colonizers of a new planet will not have to live at subsistence level, but, as we have seen in earlier chapters, they are still going to have serious supply problems, and with their place of origin 4 or 5 light-years distant, they are going to receive no help from that quarter. We must remember, however, that the original American colonies eventually came under the dominion of the land they had left. They obtained military forces to help protect them from hostile tribes and from the soldiers of neighboring colonies that owed allegiance to other European powers. When the dust had settled, the colonists found that they were subject to rule from a land 3,000 miles across the ocean. Perhaps the persecution that had originally driven them away was absent, but they were becoming subject to laws framed by a government that in no sense could be said to represent them. Planetary colonists might eventually find military forces from their home planet descending to "protect" them in a new generation of ISTs. Thus they too might become subject to rule by a government in which they had no say. We must hope that interstellar distances will protect them from this fate.

Once again, though, we are digressing from the essential theme of how planetary colonists would wish to govern themselves and how law and government might develop. Certain essential principles stand out clearly:

1. There should be freedom of speech, freedom of thought, and freedom of worship.
2. The governing body should be a properly elected and constituted assembly.
3. Because of the unique conditions in the early days it might have to exercise its mandate in a fairly authoritarian manner.
4. Though authoritarian, its rule must be sensible and just and there will have to be safeguards and methods of appeal to ensure it does not develop dictatorial tendencies. It will draw on the knowledge and advice of experts in many fields.
5. The right to private property, other than small and relatively unimportant items, must in the early days be restricted for the common good. Nevertheless, as terrestrial civilization started to spread out across the face of the planet this should certainly be relaxed. Human beings have always taken delight in having

their own property and have always responded well to incentives. Once the colonists are settled, private enterprise is likely to prove much more effective than socialism.
6. Personal ability and value to the community must be the criteria by which men and women are judged.
7. Minority interests must be safeguarded.
8. There must be a strict code of laws regarding criminal activity.

In the very early days of a planetary colony, when numbers would still be relatively low and geographical boundaries restricted, the preceeding eight points seem essentially valid, and it seems they would remain valid even after the entire planet had been populated. By the time that day arrived, the more rough-and-ready aspects would have been replaced by sophisticated forms of legislation and government. It is possible to discern the parallel between the early government of the thirteen former colonies and that of the present-day United States of America.

Let us consider the forms of government that the peoples of a "new Earth" might select once their dominion was firmly established over the entire planet.

It would seem preferable on the whole that the colonists, or more correctly their descendants, should remain one nation. The problem is that as remote regions of the planet were populated the peoples of those regions might feel they had less and less in common with those who had elected to remain closer to the original settlements. And if the authorities in the regions around these original settlements were determined to exercise their rule over the entire planet, the result could easily be trouble. Eventually there might be armed conflict. Here on Earth many political economists and philosophers have preached the merits of a United States of Earth having hegemony over the entire world and free at last from the scourge of war. It's a fine concept, but who rules and from where? A Federated States of Earth might prove easier to organize, but there still would have to be a central government and one capital city. Federation works well in countries like the United States and Australia, but these are relatively small entities compared to an entire planet.

If the original colonists were drawn from one particular nation and language, a single, planetwide state might be realizable. If, on the other hand, several nations were involved in the original mission, this is less likely. As the numbers of each nationality increased there would probably be a form of polarization based on language, social customs, and possibly also color. Geography could also play a part. Descendants of colonists from temperate regions of the Earth would in all likelihood prefer temperate regions on their new planetary abode, and descendants of those originating from the tropical regions of Earth would probably choose tropical regions of the planet. Even if all the members of the original mission had had one nationality and language, another independent colonizing mission arriving a little later might have different ones. So for better or worse, it seems likely that separate nations on the terrestrial model would eventually develop. There is nothing wrong with this. The danger lies in the possibility of one nation coveting the territory, minerals, port facilities, or raw materials of another and being prepared to use force to acquire them.

The forms of government likely to develop could well be replicas of those on Earth today. Hopefully they would be truly democratic—and remain so. Democracy may not be perfect and probably no system of government ever can be. The fact remains that it does work and it permits a high measure of something very precious—freedom!

So much for government. What of law on a planetary colony? As long as this remained a colony in the absolute sense—a relatively small number of persons occupying an equally small region—the essentials of law are fairly well covered in the eight points outlined earlier. But once the colony becomes either a single or a multinational planet there must be one, or a number, of judiciary systems. Like government, law must become more precise.

The modern world has hundreds of legal systems. Most of these are derived from principles and methods that have a common origin in a small number of groups. There is, for instance, civil law and common law. Civil law tends to embrace the experience and ideas of Roman law and is practiced in various parts of the world. Common-law systems are derived from the common law of England and are used in most English-speaking countries. To go into these in detail would be beyond the scope of this book.

Contemporary governmental and legal systems here on Earth probably are different from those that would eventually develop on a remote planetary colony. There are two reasons for this:

1. By the time such developed planetary colonies had been created, one or two centuries will have passed. By then law and government on Earth may have changed to suit changed requirements.
2. The circumstances attending a fully developed planetary colony on another world, no matter how Earth-like, would still, in many ways, be different from those obtaining on Earth. Government and law would probably reflect those differences.

Governments of terrestrial planetary colonies in early days would probably be simple elected assemblies that would have a part to play in some of the major long-term decisions for the colony but would leave the day-to-day decision making up to an efficient and authoritative executive body or council. The whole object of governing bodies must be first for the colony to survive, then slowly develop. To these essentials nearly all else would have to be subordinated.

13

Back to the Past

AN EARTH-LIKE PLANET being colonized by our descendants would, for the first couple of decades, be an odd place, for it would become, on a material level, a hybrid society. Men and women from a highly technological Earth would have brought with them the skills and knowledge of their time as well as a sampling of the fruits of that knowledge. Nonetheless, it would be physically impossible to bring all the necessary supplies and equipment to set up an immediate facsimile of the civilization they had left behind. To some extent the society would be pure 21st, 22nd, or 23rd century. In other respects it would soon border on the medieval, for the planet would be in a virgin state and the humans would literally be starting from scratch. The amount of high-technology material they could have brought will be small. The engineers in their midst would have the expertise to build great bridges, but they certainly would not have the steel. When one surveys a bridge like the Golden Gate at the entrance to San Francisco Bay, it is clear that the materials for such a structure or the equipment to place the components in position would have to be produced indigenously. The steel would still be in the ground as iron ore. Mining engineers would know how to extract it, mineralogists and metallurgists would know how to treat it, but knowing would not be enough. The mines, the iron-smelting works, and the steel-rolling mills would not be there. To create them probably would take a lot of time. The colonists' first priorities are going to be shelter and food. Land will have to be cleared and housing erected. These structures will probably be built from easily and indigenously obtained materials such as wood, stone, and clay. A substantial number of prefabricated buildings could be brought from Earth, but there will be a limit to their numbers. Since the first new buildings must entail the use of local resources, it will be of interest to discuss these.

The planet, being Earth-like, will contain rocks, mud, and clay, and these are the materials, along with timber, that the colonists will use extensively. A great deal can be done with these materials, which human beings have been using since the dawn of history. The ancient Romans, for example, had no knowledge of modern technology, yet they possessed a most excellent water-supply system capable of providing Rome with 220 million gallons of water a day. The Romans also had residences warmed by hot air flowing in channels under floors, an efficient sewage system, and thousands of miles of splendid roads with some fine bridges and great aqueducts. All this was done with the use of entirely local resources within the limits imposed by a not terribly advanced technology and with a total lack of mechanical aid. It is clear that planetary colonists, a large proportion of whom would be highly skilled in the scientific and technological disciplines of future centuries, could achieve all this and more from local resources. Certainly they would have great earth-moving machines, even if in limited numbers. However, sooner or later these machines would be rendered useless for want of spare parts. In time, as raw materials became available and an industrial base was built, they could be brought back into service or replicas constructed. Thus there is every possibility that sometime during their first decade on the planet the colonists would be forced to revert to the techniques of a much earlier age on Earth. It would be eminently sensible to include in their archives detailed records of these techniques.

The production of food crops will be of paramount importance to first-generation colonists. Food crops require water, and if rainfall is not always adequate, irrigation becomes essential. Construction of channels and canals to carry water by the colonists would be easy at first with their great powered diggers, but if these eventually proved inoperative for want of fuel and/or spare parts, what then? Machines of this nature are ruggedly built and will stand up to a great deal of harsh treatment in all climates. Nevertheless, it is by no means inconceivable that all machinery brought from Earth would become worn out long before a transplanted technology was available to replace it. Perhaps for a generation or more, the colonists could find themselves compelled to adopt the simpler techniques and practices of their forebears on Earth. Therefore we will discuss some of these practices.

Since the earliest times much of the water in Earth's rivers has been running to waste, though a number of these rivers were utilized as a means of slow, cheap, bulk transport. Early canals and reservoirs were dug by hand, animals being used to remove the soil and sand. The banks were waterproofed by the only impermeable material available in sufficient quantity—clay. Canals such as these were used over considerable periods of time, during which they served their purpose well, though doubtless with frequent repairs to waterproofed banks. The main purpose (perhaps the only one) of such canals was irrigation. Since we are envisaging a planet similar to Earth in most respects, it is reasonable to assume that it will possess rivers. It will also quite likely have arid and semi-arid regions. By the construction of irrigation canals or channels drawing water from rivers and their tributaries, these regions could be rendered wholly fertile. It is possible that the colonists will adopt a technique of this sort.

We must now address ourselves to another fundamental matter and one equally vital—energy. No doubt nuclear reactors could be brought from Earth and eventually assembled on the surface of the planet, but only a very limited number could be brought. As a new terrestrial civilization spread farther and farther from the initial base or bases, the problem of power would be increasingly felt. It is most unlikely that indigenous stocks of nuclear fuel would be immediately available. Stocks of oil and gasoline brought from Earth might last for a time but not indefinitely—assuming that the smelly, inefficient, carbon-dioxide-producing internal-combustion engine was still in vogue. However, vast, untapped reserves of natural petroleum might exist in various parts of the planet. Natural petroleum derives from the anaerobic decay of very small marine creatures under special conditions. It is thus, in effect, a product of ancient seas. However, unless the migration of the oil is prevented, it flows upward to the surface, where the liquid components evaporate, leaving behind solid waxes and asphalts. It is essential that if oil is to be commercially exploitable, some geological feature (known to oil men as a "trap") must exist to prevent upward movement of the liquid hydrocarbons. This can take the form of a structural trap, formed by faulting or folding of the rocks. The oil reservoir forms within the anticline or upward fold, and it is essential that the oil-containing rock, generally a sandstone because of its porous nature, lie between impermeable layers of rock. Lateral migration of the oil is prevented by the shape of the anticlinal fold. If there were extensive oil deposits on the planet these would be found and tapped, and then the crude oil would be taken to "cracking" plants for fractional distillation into its various components. This sort of procedure would require considerable time and effort, and before it can be accomplished most internal-combustion engines are going to be rendered impotent for lack of suitable fuel. Some might continue to function after a fashion on methyl or wood alcohol, which can be more easily produced.

The colonists could, of course, revert to using an old and well-proven source of energy—steam. To generate steam requires fuel and, of course, water. In the absence of oil, wood or coal could be used. Despite its relatively low calorific value, wood is perfectly acceptable. The early American railroads, which did so much to open up the West, worked well for years on this readily available fuel. Coal is another matter. Most is deep-mined. Deep mines (and most are really deep) take time to develop, and on the planet being colonized there would be no mechanical coal cutters. Modern coal mines on Earth have extensive underground rail tracks or conveyer belts to move the coal. On the colonists' planet shafts and galleries would have to be created the hard way—by pick, shovel, and human muscle. Even given supplies of wood and coal, iron mines and steel plants would be required to fashion steam engines—and these require fuel (generally coke made from coking coal). So back to stage one.

If nuclear energy proves inadequate, hydrocarbon fuels not yet attainable, and coal still impossible to extract, other possibilities remain. One is the solar power that we are already developing here on Earth. The other is hydroelectric power, though this is dependent on extensive catchment areas at sufficient altitudes, and although the colonists' civil engineers would certainly possess the knowledge to create great dams, they would lack the material means to do so. The same problem would exist for mechanical engineers who would wish to create turbines, generators, and power transmission lines. All these problems could be rectified in time, but there is an old proverb that assures us that "as the grass grows the cow starves."

Hydrogen, a useful if dangerous fuel, can be produced easily and in reasonable quantity by the electrolysis of water. But this would require a steady supply of electric power. What then can we think of next in the context of energy? The answer might be methane, which on Earth often occurs naturally in deep coal mines (often with catastrophic results) and in swamps. But can it be produced simply and at will? The answer is yes, although the method might not seem particularly pleasant.

The colonizing mission will have taken with it a certain quantity of livestock—most probably sheep and cattle—in the hope that on the planet they could survive and multiply. If all goes well, therefore, the colonist farmers will find themselves with quantities of sheep and cattle dung. Dried cattle dung is a widely used form of fuel in some of the poorer countries on Earth, India being a classic example. Used in this manner its value as a fertilizer is lost (what is not lost, unfortunately, is the odor). By subjecting dung to anaerobic fermentation (fermentation in the absence of oxygen), methane is produced. The waste slurry resulting from the process has a higher nitrogen content than that in untreated dung; it therefore contributes a good fertilizer for the soil. With simple materials brought from Earth in quantity it would be possible to construct cheap methane generators of around 140 cubic feet capacity having provision for continuous recharging, each producing a steady output of methane gas. Methane is the simplest of the hydrocarbons, with the formula CH_4. On combustion as a source of heat or power it produces carbon dioxide and water in the same way as the more complex hydrocarbons.

Having secured a moderate supply of methane, could it be used in conventional internal-combustion engines? The answer is yes, though the engines would require modification. The Humphrey pump, materials for which could have been brought in quantity from Earth, is an easily constructed internal-combustion unit capable of pumping water. The water being pumped replaces both the piston and the flywheel of a conventional power unit and requires no separate pump. The pump has a relatively low efficiency factor—about 20 percent—but it can use almost any gaseous fuel, including methane. It is cheap to produce and easy to maintain. Methane could also be used as a fuel in simple steam plants that could power dynamos, thereby producing a supply of electricity, some of which might be used in the electrolysis of water to produce hydrogen. We must remember, however, that all this depends on the uninterrupted intestinal output of livestock!

On parts of the planet there could be fast-flowing streams, though the horsepower produced by these would be inadequate for the hydrogeneration of electricity, even if the facilities were available. However, very limited power could be produced by a simple waterwheel apparatus. This is capable of transmitting power almost a mile by a reciprocating wire-power transmission system. The waterwheel turns a crank, raising and lowering the weighted corner of a triangular frame, which pivots on another corner to transfer the motion to a horizontal wire. This moving wire is supported at intervals by chains suspended from poles. At the far end of this wire a second frame transfers the motion to a vertical reciprocating pump. The overall crudity of the apparatus cannot be denied, but, like so many crude arrangements, the device has been shown to function well. An added advantage is that there is little to go wrong, and if anything should, it can easily be repaired.

To envisage the use of such crude contrivances by people able to transfer themselves to a world of another star may seem ridiculous. But is it? A few years ago a television program drew a parallel with what we have been suggesting. A worldwide epidemic resulted in a mortality rate so high that only a handful of people survived. It was soon apparent to them that, though surrounded by power stations, factories, airports, and railroads, they could not make use of these marvels until the world's population had expanded to something like its former extent. The survivors attempted to preserve technical know-how for future generations, but their own life-style became largely agricultural. If there was no fuel to power tractors there was still food to "power" horses. The parallel with the early days of a terrestrial colony on a distant planet is not an exact one, but many of the same basic problems are recognizable.

We have digressed somewhat from the matter of water supply. One possible solution might be the provision of a large number of rainwater "tanks" sited where the need for water is greatest. These "tanks" could be large, possibly rectangular pits dug in the ground and lined with an impermeable material such as polythene sheeting, assuming supplies of this were available in sufficient quantity. This could be followed by a layer of cake mud or clay and then one of cement mortar. Some of the water in the finished "tanks" would, of course, be lost by evaporation, but in a region of adequate rainfall this loss would easily be made good. Such stored water would need to be filtered, and sediment, which would quickly block the filters, would have to be removed. This is normally achieved on Earth by large sedimentation tanks. However, these call for considerable quantities of concrete, and although concrete could no doubt be prepared from indigenous materials on the planet's surface, its production normally involves fairly sophisticated plants and a considerable degree of heat. A more easily obtainable material might have to be employed. The process of filtration (on a small community basis) could be achieved by a simple concrete-lined tank having a sloping bottom. This would be provided with an inlet baffle to prevent incoming water from stirring or disturbing the water already in the tank, a scumboard or barrier to prevent egress of floating matter, and a sludge drain at the lowest point in the tank. The filtration process (essential in the case of drinking water) could be achieved by a relatively easy-to-construct sand filter in a concrete-lined tank.

We have also mentioned the possibilities of seawater, presumably in abundance on an Earth-like planet. At that point we were discussing ultramodern desalination plants. Here we will consider the less-sophisticated techniques the colonists would employ initially. The Sun's heat was used to desalinate seawater on Earth in the past. This is a method best suited to hot, dry regions where the star's rays are strongest and where cloud cover is generally absent. The technique is simple: Seawater is led into shallow channels under glass (we must trust the colonists have reasonable supplies of glass). Water evaporates and then condenses on the underside of these appropriately placed glass panels. The panels are sloped and the condensed water runs down their lower surface and runs steadily into fresh-water channels that lead to closed storage tanks. Crude yet strangely efficient. It is probably superfluous to remark that if such a simple system is to provide fresh water in any quantity the "evaporation area" under glass must be extensive. Such water comes in the category of distilled water, which, since it contains no dissolved air (or anything else, for that matter), is not particularly pleasant to drink. It has a very "flat" taste but it could be transformed into reasonably potable water were air to be pumped continuously through the water in the tanks.

An Earth-like planet would also be expected to contain large and small fresh-water lakes. The water from such lakes might be almost immediately drinkable, but preliminary tests would, of course, be made. Streams whose courses had traversed territory containing lead ores would probably, if they ran into a lake, deposit a small amount of dissolved lead salts. The stream itself could easily have a lead content above the acceptable minimum, but the lead salts would be well diluted when the stream flowed into the lake. However, if a number of streams had been pouring lead-containing water into the lake over a number of years, the lake could build up a high lead content. Lead is likely to be the principal offender, but stream water could conceivably contain minute but potentially harmful amounts of cadmium and/or mercuric salts, depending on the regions traversed by the stream. Cadmium occurs principally with zinc in the mineral zinc blend, though the ore rarely contains more than 0.5 percent cadmium. Mercury, the chief ore of which is cinnabar (mercuric sulphide), tends to occur (at least as far as Earth is concerned) along lines of profound volcanic disturbances, but streams might carry strong traces over a considerable distance. Any buildup of mercuric compounds in the human body is extremely harmful, especially to the brain. On an Earth-type planet it is not unreasonable to suppose that these elements, especially lead, would be present as ores in the planet's surface. The colonists therefore would be wise to have stream and lake water tested. It is a strange irony of present-day life that many, if not most, countries continue to permit the addition of tetraethyl lead to gasoline. The consequence of this superlative piece of folly is that millions are exposed to the menace of lead assimilation. It is to the credit of countries such as the United States that the use of lead-free gasoline is becoming mandatory.

The effect of organic detritus in lake water also must be considered, though risks from this are likely to be less than from an overconcentration of soluble metallic salts. But we must remind ourselves that extensive testing facilities might not be available. A settlement far from the original landing and some years after the initial arrival might find itself reduced to a more basic form of life. It will simply have to do the best it can and take what chances it must.

Let us now consider another essential in the overall context of this chapter—shelter and buildings. At first buildings would almost certainly be prefabricated structures shipped from Earth. These could be elaborate, large, and enduring, but the quantities would be limited. As the colonists married and multiplied, new shelters would have to be built for them. The larger and more elaborate prefabricated structures would become the first hospitals, laboratories, schools, and administrative centers on the planet. Those for families and for individuals would be simpler, smaller, and probably Spartan and stark. Color TV, refrigerators, and other symbols of our affluent society would almost certainly be absent—and perhaps in the early days that would be no bad thing. It would be desirable for the colonists to be a tough and hardy breed, and the presence of such luxuries might have retrograde effect on the development of these essential qualities.

New sources of shelter would be produced from indigenous materials. Not for many generations (not indeed until technology is catching up with the technology in vogue when the colonists quit Earth) will great structures of steel, glass, and concrete arise. The buildings of the early colonists would be simple and functional. This does not mean a reversion to medieval constructions that made no provision for modern hygiene.

Let us look briefly at building techniques on Earth in the past, as well as today among the poorer and less well developed nations. The parallel between these and what may be employed by future colonists from Earth may not be exact; parallels rarely are. There could, however, be a remarkable degree of similarity.

On the chosen planet, just as here on Earth, man's most basic need, after food and clothing, would be shelter from the elements. Nomadic peoples in the past seem to have done remarkably well without fixed abodes, but these peoples generally lived in warm lands. In more temperate climates during high summer, thoughts of a blissful, carefree, nomadic existence exert a profound appeal. We can dream of wandering through forest glades, beside gurgling hillside streams, lying under hedgerows ablaze with wild flowers, and sleeping under star-powdered skies. It sounds idyllic and for a brief period might prove so. But then would come autumn with its winds and cold, driving rain and winter with its frost, its blizzards, and its stark, leafless trees. The dream soon fades. We need shelter. It is essential for survival. Primitive man satisfied this by dwelling in caves. His descendants then began to construct buildings—if crude huts are deserving of the term. Key factors in the design of these were the prevailing climate and the materials immediately available. In hot, arid climates, dwellings having very small openings in their thick walls of dried mud were built to keep out as much of the heat and sunlight as possible, whereas in damp, rainy climates the normal practice was to construct sloping roofs of grass or of rushes so that the rain ran off without penetrating the interior. In areas where earthquakes were a regular feature, dwellings were built of extremely light material so that if they collapsed, the occupants would be less likely to sustain serious injury.

It would be reasonable to assume that terrestrial colonists, despite the temporary hiatus in their technological achievements, would, on the whole, be able to contrive something a little more sophisticated than the types of dwellings just mentioned. From skyscraper to mud hut via star ship would indeed be a cruel irony!

Naturally, on an Earth-like planet we would anticipate trees.

To what extent these trees would resemble those of Earth is impossible to say, but chances are there would be a strong similarity. The colonists would no doubt begin to process the trees for timber with mechanical saws. If these were worn out or became too costly in terms of the energy required, they would still have axes, handsaws, and human muscle. Hauling the felled timber from the forests would be more difficult. Animal power in the form of either horses descended from those brought from Earth or beasts of burden native to the planet would probably be the answer. It might also be possible to float the tree trunks down a river if a river was near. However, though it may be easy to launch tree trunks into a river, getting them out again at the point where they are required would, in the circumstances, be less than easy. Much ingenuity would be called for.

A timber dwelling, despite an unfortunate propensity to catch fire, can make a very comfortable residence if designed and built with a little forethought. Here then is a satisfactory material that the ever-expanding colony from Earth could utilize to its advantage. We are not thinking of the old type of log cabin, though as history shows, these made good shelters.

The strength of timber varies depending on whether this is measured along or across the grain. The strength can be rendered much more uniform if strips of timber with the grain running in various directions are bonded together by suitable adhesive compounds. If stocks of these were to run out before it became possible to manufacture more, it might be possible by a process of trial and error to produce a satisfactory, simple substitute from indigenous sources such as resin from trees, or bitumastic materials from carbon-rich areas.

Eventually timber rots, especially in damp climates. The application of preservatives such as creosote (especially if applied under pressure) inhibits this process. Application under pressure may not always be possible, in which case conventional coating methods will have to be employed. If stocks of creosote, phenols, and other recognized wood preservatives were exhausted, it might be possible for chemists to produce similar compounds from planetary materials.

With timber present in quantity we can rest assured that stone will abound. Unfortunately, even if it is freely available, it must still be extracted from the ground.* This will call for the use of high explosives, but we can be certain that whatever explosives were brought from Earth will be quickly consumed. Great care will have to have been taken to ensure their safe storage in the IST. Modern nitroglycerin high explosives become highly dangerous if stored for protracted periods under adverse conditions. The most common problem occurs when the highly sensitive (indeed, almost temperamental) nitroglycerin exudes from the absorbent material that, until that time, had rendered it comparatively safe. Explosives in this condition are known as "weeping" explosives and should be destroyed by experts as quickly as possible.

The manufacture of modern, nitroglycerin high explosives is not particularly difficult on Earth, where the requisite raw materials—glycerin, nitric acid, sulphuric acid, and diatomaceous earths (or other absorbent materials)—are readily available. (It must be stressed, however, that the manufacture of high explosives is not without risks. Anyone as unwise as to attempt the manufacture of nitroglycerin on a small scale at home is unlikely to see the completion of the experiment, though most of the surrounding neighborhood would certainly be aware of it!)

On another planet the materials for manufacturing nitroglycerin might not all be found, but it is conceivable that in their absence colonists could manufacture an earlier form of explosive, popularly termed gunpowder but known to explosives technolo-

---

*Once the stone and rock have been blasted they must also be moved to where they are required. Powerful high-capacity mechanized transport and loading gear would be very useful, but it is doubtful if the colonists will be able to bring such equipment with them. There must be an answer, for the ancient Egyptians had neither high explosives nor mechanical transport yet they built the pyramids. The Incas in Peru raised massive structures of stone blocks that fit together so perfectly and exactly that it is virtually impossible to insert even a razor blade between the blocks. But how they performed these remarkable feats has long been a matter of considerable debate. In the case of the pyramids it is known that the great blocks of stone were somehow quarried then shaped (in itself no mean task) and floated down the River Nile on specially constructed rafts to the site of the pyramids. Thereafter an inclined plane was probably constructed and up this the blocks, mounted on crude rollers, were pulled by thousands of slaves. However, since the colonists would not have slave labor available to them, and hopefully would not wish to have it, some other method of transporting stone will have to be found.

gists as "black powder." It is still used today and is essentially an intimately ground mixture of sulphur, potassium nitrate, and powdered charcoal in the appropriate proportions. It is fairly easy to make but has a nasty habit of detonating during the final stages of manufacture. It was the first and, for several centuries thereafter, the most effective explosive propellant. It was known to the ancient Chinese and is believed to have been introduced into Europe by the Arabs. The manufacture is not merely a matter of simple mixing, though if the proportions are right it will burn fiercely and might detonate.

After mixing, the black powder is moistened to prevent spontaneous ignition, the resultant paste being milled to reduce the particle size. The "cake" that forms after drying is broken into grains of varying sizes. Large grains are slow-burning (the term "slow" in this context is a very relative one). This characteristic produces the "slow," heaving effect so useful in quarrying rock.

The manufacture of black powder would probably lie within the capacity of a developing terrestrial colony. Sulphur is a common element on Earth. There are immense deposits of it in Texas as well as in Mexico and Sicily. Quantities of the element also can be found in volcanic regions. Charcoal can easily be made by burning suitable types of wood and powdering the remains. Potassium nitrate also is fairly easy to come by. For example, if sodium nitrate in large quantity (such as the vast sodium nitrate deposits in Chile) is available, it is possible to convert it to potassium nitrate by the action of potassium chloride in a fairly straightforward chemical reaction. But would all these materials be easily available on a planet that was not Earth but merely Earth-like? Since such a planet would have evolved from a primordial nebula in much the way that Earth did, the answer is a guarded yes. The elements that constitute it are to all intents and purposes going to be those that went to make up Earth. Sulphur is a common element on Earth and so also should be on a distant facsimile of our world. Potassium nitrate, a compound of potassium, nitrogen, and oxygen, all three of which are found in great quantities on Earth (especially nitrogen and oxygen, which are the principal constituents of our atmosphere), is hardly likely to present a problem. The same could be said in respect of potassium chloride. Thick deposits of this compound result from the evaporation of ancient seabeds. One of the best examples is the huge deposit near Stassfurt in Germany. Naturally occurring sodium nitrate can be used in lieu of potassium nitrate in black-powder manufacture. The fact that sodium nitrate is hygroscopic (it readily takes up and retains moisture) renders it much less suitable, however.

Here then we have an explosive capable of being produced with relative ease (and a certain degree of hazard), which happens to be particularly well suited to quarrying operations. Colonists should be able to blast convenient rock faces to secure stone for building. Shaping that stone into blocks will be less easy, and assuming no technological aids are at hand, the old-fashioned hammers and chisels will be used. It can be done by primitive means—the Incas did it.

Science fiction visualizes disintegration beams that would make short work of so mundane a thing as rock quarrying. In the early days of a terrestrial colony that might be precisely the method employed. But as the power sources for these beams became exhausted, with no prospect of quick replacement, circumstances would change radically.

Bricks represent another building possibility for colonists. Bricks are made from clay and shale and are "burned" in a kiln. Producing these should be within the colonists' powers. We might also consider concrete. This is regarded as a 20th-century construction material, but in fact a form of concrete was used by the Romans in their grandiose aqueducts, amphitheaters, and baths. The remains exist to this day. The production of a modern grade of concrete would, however, be beyond the powers of terrestrial colonists until they redeveloped an adequate technology.

Once the initial resources brought from Earth are depleted, the colonists will be compelled to exist in circumstances quite different from what they knew on their home planet. We do not know what technical powers will have been developed between now and the time a colonizing expedition sets out for a new home among the stars, and it is conceivable that the colonists might have less of this "make-do-and-mend" existence than we have been imagining for them. Nevertheless, it seems clear that a severely reduced standard of living will be endured for some years after the resources and materials brought from Earth run out. We are thinking in terms of *colonizing* a planet, of setting up on a world like Earth, a total extension of the civilization and technology of the home planet. It would be physically impossible to transport every item essential for such a vast undertaking no matter how many ISTs were involved. The catching-up process, however, would be on a much shorter time scale than that which produced 20th-century technology. Modern terrestrial technology is largely a child of the past 100 years. Time was needed to hypothesize, experiment, discover, and test. Colonists would already possess the knowledge and skills appropriate to the time when they left Earth. These they would use as well as pass on to future generations, so long as "Nova Terra" was well endowed with natural resources. The catching-up process could be surprisingly quick. It could not be immediate. There would have to be a transitional period that might be less than pleasant.

14

Moon Colony

THROUGHOUT THIS BOOK we have visualized a real terrestrial planetary colony as possible only on a world showing a strong resemblance and affinity to Earth. By "colony" we have meant a world where the human race can take up the reins of terrestrial life again, where it can extend and develop and eventually produce a civilization approximating that of Earth, though hopefully (and perhaps too optimistically) free from its worst sins and shortcomings. Such a world can be found only orbiting some other star, and it will, therefore, be at such a distance from Earth as to ensure all the difficulties of isolation we have described.

But what, the reader is surely entitled to ask, about some of our sister worlds about the Sun? In comparison with the worlds of other stars, the distances involved are trifling, and help would be available from Earth within a reasonable time. Unfortunately, the planets of the Solar System, with the exception of Earth, are totally unsuited for terrestrial colonization—colonization, that is, on the extensive and permanent scale we have been considering on these pages. The planets Jupiter, Saturn, Uranus, and Neptune are all huge "gas giants" and must be eliminated from the reckoning at once. Pluto, the outermost planet, is too distant and too cold, and we are not even certain of its true physical state—it could be little more than a methane snowball. This leaves Mercury, Venus, Mars, and, of course, our own familiar Moon. Mercury can quickly and easily be disposed of, for it is merely a small, extensively cratered planet, totally devoid of atmosphere and lying only about 33 million miles from the Sun—much too close for comfort. Venus can follow it. Venus is a planet comparable to the Earth in dimensions but far too hot and having a climate of carbon dioxide and dilute sulphuric acid rain. Only the Moon and Mars remain within the bounds of possibility, so let us look a little more closely at these. We will start with the Moon—the closest celestial body to Earth and the best known. It is a world we can see, if skies are clear, for most of each month. It is a mere astronomical stone's throw away. What could be more fitting than a colony on the body so often and eloquently described as the "Queen of the Night"? The answer is probably nothing, except for the trifling fact that there is no air to breathe and gravity is only one-sixth that of Earth. Lunar colonists might gradually adapt themselves to the latter—but hardly to the former! Thus a colony on the Moon, if it is possible, must expect to live under very different conditions from those obtaining on Earth. It would seem that sojourns on the Moon would be transient affairs. Nevertheless, several science-fiction writers have put colonists on the Moon, even going so far as to envisage some who were born there and have come to regard it as their permanent home. This seems less than likely. To live for a lifetime under totally artificial conditions when Earth, with all its winds and waves, its grasses and trees, is only a quarter of a million miles removed would surely be a strange existence. And should there be Earth people who wish to free themselves from the tyrannies, machinations, and double-dealings of terrestrial politicians, dictators, and commissars, they are going to have to move very much farther than that from their native world.

The Moon is a much smaller body than the Earth, with a diameter of only 2,160 miles compared to the 8,000 or so miles of Earth. The Moon has a mass of only 1/81 that of Earth, which explains why people on the Moon would have only one-sixth their respective weights on Earth. Walking on the Moon is not too difficult, as the United States astronauts were quick to demonstrate. Prolonged exertion could, however, prove very tiring.

It was firmly believed at one time that, due to the Moon's virtual lack of atmosphere, meteoric bombardment would render it essential to construct bases beneath the surface. Now it does not seem that this is so.

The first true pioneer base on the Moon probably will be a group of self-contained cylindrical structures, although there might be a brief period prior to that in which a grounded space vehicle (or vehicles) hitherto in orbit around the Moon would be used. We might look for a base of this nature within the next couple of decades, probably before the end of this century. It is most unlikely, however, that such a base would be permanently manned, though it probably could sustain a number of persons for weeks or even for a month or two. Clearly this would hardly represent a colony in the real sense of the word.

Nevertheless, during the next century (and probably within its early to middle decades) a colony of sorts could be established on the Moon. However, the highly artificial conditions that must always prevail on that utterly barren world will strictly limit the number of people who can live there. In the beginnings it is reasonable to assume that most personnel, both male and female, would be there largely in a scientific capacity but with a small element of administrative and ancillary workers. Toward the end of this chapter we will look at the highly speculative concept of "moon cities." In the meantime, let us return to the early decades of the next century and see what we may justifiably expect. By then we shall assume that a permanent lunar base has been established. This comprises chiefly a number of domes, some larger than the others and all connected by pressurized passages or walkways. In the dark, star-strewn sky above hangs a "full" Earth showing its complex pattern of oceans, continents, and clouds. Some areas of the Moon are brilliantly lit, while others are in the deepest shadow. The photographic expression "soot and whitewash" will have a lot of relevance on the Moon when the sunlight brilliantly illuminates plains but throws crater walls and mountain ranges into strong relief.

The largest dome might consist of several stories and contain living quarters. The most stringent safety precautions would have to be taken here (as in every other area of the lunar base) at all times. The lunar landscape has a strange, almost fantastic beauty, but it is, nonetheless, the realm of death. Failure of pressurization in a building, or part of one, would expose the human bodies within to vacuum—and the human body does not appreciate it! If this happens, the blood will boil and the body explode. It would be unpleasant and messy. And that fate awaiting the unwary defines for us precisely the difference between a "colonist" on the Moon and one on an Earth-like world, even if it be 10 light-years distant.

The other domes on the Moon probably would comprise assembly halls, lecture rooms, storage facilities, laboratories, and an astronomical observatory. Around them we might expect considerable evidence of hydroponic farming. As was explained in Chapter 10, in this sort of farming the plants are nourished by nutrient liquids continually circling beneath them. Hydroponic techniques could be very important on the Moon, for such "farm" products would cut down on the amount of material to be brought from Earth and would therefore help the colonists to be a little more self-reliant.

The people on the Moon would certainly have some form of mechanical transport—we might term it a "Moon crawler"—to explore and collect geological samples from the immediately surrounding area. Such a vehicle would have to be electrically propelled, since internal-combustion engines will not run in the absence of atmospheric oxygen. These "crawlers" would be crewed by personnel in pressure suits, as in Apollo 15. The day when lunar transports are sufficiently enclosed and robust to allow individual pressurization must be regarded as distant.

Why attempt to place a colony, even a transient one, on the Moon? Certainly there are scientific reasons, and one of the principal ones would be geological (or, more precisely, selenological) research. The lunar atmosphere, assuming there ever was one, departed permanently into space aeons ago. Since then there has been no erosion, such as occurs on Earth due to rain, frost, wind, and ice. Therefore, the surface of the Moon, apart from some thermal disintegration due to severe alternate heating and cooling and the formation of a few more recent small craters, is virtually as it always was—a totally unchanged world.

On the Moon it is unlikely that erstwhile colonists would be in any danger from "moonquakes," and the threat of meteors would appear to be overrated. However, the lunar surface is exposed to all the various types of radiation emanating from space, and methods of protection from this harmful radiation would have to be devised. In Chapter 11 we saw how Earth's upper atmosphere and ozone layer protect us from ultraviolet and X rays. On the

Moon there is no atmosphere at all, and these radiations come flooding in totally unimpeded.

The lack of an atmosphere is the greatest single disadvantage to life on the Moon—no air to breathe, no atmosphere to protect. To complicate matters a little more, we must remember that the lack of an atmosphere means sound waves cannot be propagated. Radio is the only means of communication between space-suited colonists out in the open. Moreover, the absence of an ionosphere rules out all possibility of long-distance short-wave transmission and reception, though the presence of orbiting communications satellites would obviate this difficulty.

One overwhelming advantage enjoyed by lunar colonists over fellow terrestrials on distant Earth-like planets of other suns would be the quickness and ease of communication with the home planet. At a quarter of a million miles, the Moon is only a light-second distant.

There is little doubt that life in a lunar colony would be full of interest and novelty, but there would always be danger and difficulties and problems of the most bizarre kind. The cost of security would be eternal vigilance. The risk of a leak in one of the structures of the lunar base would be ever-present in the minds of the colonists.

Astronomers on the Moon will have a tremendous advantage over their Earth-based colleagues, for they will not be handicapped by having to peer through layers of unsteady, obscuring atmosphere. Indeed, it is from a lunar observatory that some of the planets of the nearer stars may first be detected. In that sense the way to the stars might be via the Moon! Another advantage for lunar astronomy will be the very low gravity compared to Earth. This will render it much easier to construct large telescopes, especially those of the radio type that are exceptionally cumbersome.

What will be less welcome to scientists (and to everyone else in a lunar environment) are the almost terrifying temperature extremes. At lunar noon this is +210°F (almost the temperature of boiling water at sea level), and at lunar midnight it is –250°F. Since +32°F represents the freezing point of water, lunar midnight is clearly chilly! The maximum temperature given is that found in the lunar equatorial regions. Daytime heat at the polar regions is, as might be expected, less, though by no stretch of the imagination could it be described as anything but exceedingly hot, and the minimum at the poles is, of course, even colder. It is also necessary to remember that daytime on the Moon lasts not for 24 hours but for 14 of our days. Lunar nighttime also lasts for 14 of our days. Clearly the Moon is a place of extremes. Temperature changes of this range and magnitude will involve considerable care in the choice of construction materials, especially those used for the domes and walkways. If a pressure dome cracks due to differential expansion, the results would be decidedly unpleasant.

Could the Moon ever become a terrestrial colony in a more nonscientific sense? Some think this might happen in time, and one or two writers even go so far as to see it as a future super holiday center for tourists from Earth. There is no harm in looking at a few of the possibilities as long as we remember that colonization in the fullest and widest sense of the word is most unlikely ever to be a possibility on this airless, barren, and dangerous world.

It would certainly be possible to build small cities on the Moon if the entire complex were hermetically enveloped in a great pressurized dome. The technical difficulties are obvious. This dome must not leak. It must be invulnerable to breakage or fracture by meteors, by heat and cold, and by any other external agency. It might easily resist micrometeors, but what of an occasional moderate-sized chunk of stone or iron reaching the Moon's surface from interplanetary space? Every so often, meteorites that are too big to burn up in the relatively thick atmosphere of Earth reach the surface of our world. The result is often highly spectacular even though, as far as is known, no one has yet been killed or injured by one of these celestial projectiles. What would a comparable object achieve if it struck the pressure dome of a lunar city? The result is horrible to contemplate. At the present time it is difficult to imagine a plastic that could resist such an impact. And, of course, the dome would also have to withstand a lunar day temperature of +210°F and a nighttime low of -250°F. No plastic might suffice and recourse might have to be made to some stronger, wholly opaque material. Even such a material would have a breaking point, however, and a sufficiently large meteor might crack its surface.

Low gravity would prove awkward to new arrivals from Earth, but in a fairly short period, they probably would accustom themselves to this. The Apollo astronauts, after only a few days, seemed to be doing all right. Children born on the Moon, despite acquisition of terrestrial muscles by inheritance, would not need to use these as fully as they would on Earth. With the passing of years, this might result in the partial atrophy of the relevant muscles. If any of these Moon-born terrestrials eventually went to Earth we must assume they would experience muscular problems. They would probably be tempted to take the next ship back to the Moon.

We must also consider the effect of ultraviolet and X rays on human beings living within a lunar city covered by a transparent dome. The material must not permit passage of these harmful radiations; if it did, the occupants would soon be very sick. Since the radiation must be filtered out, this process might be incorporated with a scenic effect suggested by one or two science-fiction writers wherein the illusion of terrestrial clouds would be created in the upper reaches of the dome. Visitors from Earth who might regard this as retrograde since they have come, however briefly, to experience lunar conditions and live under lunar skies could be regaled with short trips outside the city in "Moon crawlers" specially pressurized for their convenience.

But what of transportation on a larger, more commercial scale between "Moon cities"? Admittedly this is looking far ahead into a decidedly problematical future. Rocket-propelled craft would seem the most logical answer, though some theorists foresee a day when underground pneumatic or electrically driven and pressurized tube trains could go from city to city. The terminals for these would be within the lunar cities, and the trains (for want of a better term) could pass directly between cities without ever emerging on the Moon's surface. Moonquakes or other lunar geological phenomena might, however, render such projects risky. Another futuristic concept that has been suggested is an elevated monorail running from city to city, even to cities and communities on the Moon's far side. Presumably if men can one day reach the stars, they can build railroads on the Moon! It is fascinating to reflect on what such a journey would be like—over desolate lunar "seas," through great valleys, around the rims of vast craters, and at last across the horizon to the region in the skies of which Earth has never and will never be seen.

We must remember that visitors to the Moon, be they permanent, transient, or tourist, will have a choice of locale, which we can term "near side" or "far side." The Moon keeps virtually the same side or face permanently turned toward Earth. Until the Soviet Union in the early 1960s sent around the Moon a probe that succeeded in sending back crude photographs of the hitherto hidden side, we on Earth had absolutely no idea what its topography was like. It was generally assumed (not unreasonably) to be similar in virtually all respects to the side with which we are so familiar, and this proved to be the case.

Those reaching the Moon may be due for a number of surprises. One will almost certainly be the distance to the lunar horizon. It will seem far greater than it really is. Because the diameter of the Moon is considerably less than that of Earth, its horizon is much closer. Oddly enough, this is not the impression given—a good example of an optical illusion. The distance to the Moon's horizon is a mere 3 kilometers (about 2 miles), a distance that in normal circumstances could be covered by a brisk walker in about 20 minutes. Other aspects of the Moon that surprised the first astronauts to reach it are the lunar mountains. In appearance and proportions these proved to be severe disappointments. For many years astronomical artists had been producing imaginary lunar landscapes with stark, jagged, steep-sided, very alien-looking mountains. They could hardly be faulted for these interpretations, since astronomers and geologists of the prespace eras had assumed that the total absence of wind, rain, frost, and snow must have left mountain ranges that were very steep compared to those of Earth. Certainly those ranges always looked exceedingly impressive in telescopic photographs taken from terrestrial observatories. This was primarily due to the long shadows the mountains cast. But these mountains were never all that high. Most are better described as gently rolling hills. Even crater ramparts are unimpressive when viewed from the lunar surface. Science-fiction author Arthur C. Clarke described the situation well in his novel *A Fall of Moondust* when he wrote, "There was not a single lunar crater whose crater ramparts soared as abruptly as the hills of San Francisco, and there were very few that could provide a serious obstacle to a determined cyclist."

15

Mars Colony

THE OVERRIDING ADVANTAGE of the Moon as a colony over an extrasolar planet lies in the Moon's extremely close proximity to Earth, but its total lack of atmosphere and generally unterrestrial features give it little potential as a colony in the full and accepted sense of the word. When we consider Mars as a possible colony it could probably be said that here we have the worst of both worlds—in the literal as well as the metaphoric sense. Admittedly it is going to be infinitely easier to reach Mars than to reach an Earth-like planet of the nearest star, Alpha Centauri (assuming such a planet exists) at a distance of 4.3 light-years (roughly 25 million million miles). At a favorable opposition (that is, when Sun, Earth, and Mars lie in a straight line, with Earth and Mars on the same side of the Sun), Mars is 35 million miles distant, but all oppositions are not as favorable as this, the distance sometimes can be around 60 million miles. When Mars and Earth are in conjunction (that is, Earth, Sun, and Mars in a straight line but with Earth and Mars on opposite sides of the Sun), the distance is quite immense. Even the 35 million miles of a close opposition represents a considerable journey. A Mars colony would not, therefore, be able to whistle up aid or urgent supplies at will and would most assuredly need to take along a fair measure of its essential requirements.

Unlike an extrasolar but Earth-like world, little or nothing would be available indigenously. Mars does have an atmosphere, but it is not one we could possibly breathe, and the presence of an atmosphere leads to violent and often protracted dust storms that rage over a large part of the planet's surface. The personnel of a Mars colony would therefore have to live under highly artificial conditions, far from immediate help, on a very cold, uncomforta-ble planet. Their sun would still be the Sun of Earth but being considerably more distant, it would not appear nearly as large in the Martian sky nor would anything like the same amount of heat reach the Martian surface. Mars is going to prove a very chilly place.

It is possible, as it was with the Moon, to envisage Mars as the site of a scientific colony whose purpose is to elicit as much data as possible concerning the planet. In the early days it is virtually certain that those going there will stay only a couple of years. This represents a lengthy period under the circumstances, but the changing positions of Mars and Earth relative to each other will not render a shuttle service between the two planets easy. Will a permanent colony ever develop on the legendary red planet? This is a very difficult question to answer. We must suppose that with advancing technology it will be possible to construct "lunar"-type cities on Mars, but whether any of our descendants would wish to become permanent residents—"Martians"—is debatable. As for the prospect of terrestrials wishing to vacation on Mars, it would be difficult to imagine what pleasure they could get from it other than the sensation of novelty and the satisfaction of boasting they had been there. The travel time alone might be enough to put them off—and just consider the expense!

Despite its inherent dangers, there is a fascinating beauty about the Moon's terrain. It is a world of intense light and deep shadow, black skies, Earth-filled in one hemisphere and star-powdered in the other. Mars, on the other hand, is to a large extent a drab, flat, boulder-strewn plain with a strange, pinkish sky during daylight hours. There are admittedly a number of highly interesting features. The vast volcano known as Olympus Mons is one, and the apparent evidence of once-extensive river systems is another. These are of considerable scientific interest but would probably not justify the discomfort and expense of visiting Mars for any but the most fanatical tourist. The Martian night would be dark, lit only by the stars and the two strange speeding little moons, Deimos and Phobos, so aptly described by

American author Edgar Rice Burroughs as "the hurtling moons of Mars."

It remains extremely difficult to envisage Mars as a terrestrial colony in the full and generally accepted sense of the word. However, it is better not to be categorical about these matters. Such is the rate of progress and of technology that a "prophet" can be left looking (and sometimes feeling) extremely foolish. Certainly a permanent, expanding Martian colony within the lifetime of most of us is, to say the least, improbable. However, just over 25 years ago a newly appointed British Astronomer Royal, on being asked his opinion of space travel, described it forcefully as "utter bilge." Within nine months the first artificial satellite was in orbit around the Earth; 12 years later, men had walked upon the Moon.

First let us look briefly at Mars in the light of current knowledge. It must be a source of considerable regret to many that there are no indigenous Martians. As a member of a generation that grew up in an era when an advanced civilization on Mars was still regarded as a possibility, this has proved a bitter pill for me. It used to be a source of infinite fascination to look up at the bright red "eye" of Mars in the night skies prior to World War II and let one's imagination have full rein. Was this not the planet with the wonderful, all-embracing network of canals? And if a society had achieved this, what other marvels might it not have achieved? Did a complex system of monorailways follow the routes of these canals? Did great futuristic cities rise where the main canals intersected? What strange craft sailed these waterways? Perhaps, could we produce more sensitive radio receiving equipment and more efficient antennas, we might hear the voices of Martians or the strange cadences of Martian music. Alas for all these dreams of yesteryear, we must now admit that the fabulous Mars of Schiaparelli, Proctor, Lowell, Wells, and Burroughs was a mirage. We will never see the octopuslike creatures envisaged by Wells (on reflection perhaps that is not a cause for regret). Neither can we hope to become acquainted with the lovely Martian princess Dejah Thoris of Burroughs (quite definitely a cause for regret). So what, then, do we have? An unpromising planet, to say the least. The word "Mars" has lost its magic ring. If we might parody a well-known Latin phrase, *Sic transit gloria Martis.*

Mars has a diameter of 4,200 miles, about half that of Earth, while Mars' mass is only one-tenth that of our planet. Mars is one of the smallest planets in the entire Solar System. Only Mercury and Pluto are smaller. With an escape velocity from it of only 3.1 miles per second, it has been unable to retain a dense atmosphere similar to our own. There is definite and incontrovertible evidence of running water there many millions of years ago, a fact that would indicate that the Martian atmosphere must then have been much denser than it is today. At the present time liquid water could not possibly exist under so low an atmospheric pressure.

The Martian year or period of revolution around the Sun is equivalent to 687 terrestrial days. Its diurnal rotation time or "day" is 24 hours, 37 minutes, so that a day on Mars and a day on Earth are very similar in this respect. The tilt of its axis (between 23 and 24 degrees) is also about the same, so the seasons also approximate ours but are much longer. And, of course, much colder, with the temperature rarely exceeding 0°C, although at the equator it can sometimes reach 18°C.

In 1971, when the U.S. space probe *Mariner 9* went into orbit around Mars, it transmitted thousands of very-high-quality pictures back to Earth. To the considerable surprise of astronomers, large areas of the planet were found to be heavily cratered and presented an almost lunarlike appearance. Canals might once have been suspected, but never craters. There was also clear evidence of giant volcanoes and great yawning canyons. Most assuredly this was not the Mars of science fiction! The greatest volcano yet discovered is the renowned Olympus Mons. It is 18 miles high, the crater alone being 45 miles across. All terrestrial volcanoes, by comparison, fade into insignificance. These same pictures showed another entirely unexpected feature of the planet—dried-up riverbeds, deserts marked by the flow of past floodwaters, and a considerable amount of other evidence indicating that fresh water in considerable quantities had once flowed on the surface of Mars. It is possible to date that flow by counting the number of impact craters that cut across the lines of these dried-up riverbeds. Obviously the craters were formed after the rivers ceased to flow and, since it is known approximately how many new craters are formed each million years or so (by virtue of comparison with craters of the Moon), it is possible to calculate roughly how much time has elapsed since the river systems dried up. The results indicate that the riverbeds dried up some hundreds of millions of years ago. Before that Mars must have been a warm, wet planet with a thick atmosphere. The early histories of Earth and Mars paralleled one another, but eventually there came a point when they deviated markedly. On Mars it is likely that photochemical reactions due to sunlight liberated light gases from the atmosphere. Probably the breakdown of ammonia, which is a compound of nitrogen and hydrogen, resulted in the escape of hydrogen into space due to the relatively weak Martian gravity. Lighter gases liberated on Earth by anomalous photochemical actions are retained by the heavier gravity. On both planets sunlight broke up water molecules in the primary atmosphere, producing oxygen. On Earth much of this combined with carbon from the rocks to form carbon dioxide, where it remained and was utilized by early plants and converted into the present oxygen-rich atmosphere. On Mars, however, even the carbon dioxide escaped, so the planet continued on its barren way.

Another intriguing fact brought to light by *Mariner 9* concerned the regions of the planet known as Hellas and Argyre. Both of these had long been visible from Earth but had always been regarded as uplands or plateaus. In fact they are just the reverse—depressed basins.

The two Viking probes each put down a lander component, one on the plain of Chryse, the other on the plain of Utopia (a singularly unfortunate title in view of its desolate nature). Both sites are rock-strewn wildernesses, though there is evidence of running water in the past. Temperatures are extremely low—around -22°F.

For a long time there was considerable controversy as to whether the Martian pole caps were composed of ice or of solid carbon dioxide. The orbiter sections of the Vikings finally established that they were composed of ice and might, in fact, be very much thicker than was once believed. Water is also known to exist in the surface rocks in chemically combined state. Therefore, Mars cannot be regarded as an arid waste in the full sense of the word, though there are no oceans, seas, lakes, rivers—nor canals.

Whether Mars is to become a small colony of Earth or merely a site for scientific research, a base will first have to be established there. A journey to the planet, even if nuclear-powered rockets were used, is going to take several weeks. The first men from Earth are going to find that because of the extreme thinness of the atmosphere, fully pressurized suits will have to be used at all times when walking on the planet's surface. The idea of using simple oxygen masks must be discarded once and for all. At least the alternation of day and night will not seem strange to terrestrials for, as mentioned earlier, the Martian day is only about half an hour longer than the day on Earth.

If a decision were made to set up permanent colonies on Mars, many problems would have to be faced. Such natural resources as there are on the planet—and there may not be many—will, of course, be utilized. Water obviously would be of prime importance. It has been suggested that appreciable quantities of ice might exist below the surface, but at present this is only conjecture. If it did exist, it could be extracted and melted for water. The melting would have to be carried out under pressurized conditions, otherwise all the water produced would evaporate very quickly. If indigenous water were not available, then all water would have to be transported from Earth. This alone would call into doubt the practical feasibility of a Martian colony. Even the minimum distance between Earth and Mars of 35 million miles is a very long haul for fleets of water-tankers. Oxygen, of course, would be even more vital than water. If oxygen could not be produced on Mars then it too would have to be brought from Earth. It would seem that planetary colonies within the Solar System would always be tremendously reliant upon Earth.

The surface gravity of Mars is only about a third that of Earth. On the Moon it is only a sixth. Lowered gravity is hardly likely to prove harmful—rather the reverse, and to persons suffering from cardiac ailments, it could be advantageous. It would nevertheless prove awkward until a person became thoroughly accustomed to it.

If a Mars colony is eventually created, presumably children

will be born there, and it is feasible that they will grow up so used to the lower Martian gravity that on going to Earth they will be unable to adapt to the stronger pull of the home planet. A stage might be reached when they had, to all intents and purposes, become "Martians" and could never hope to visit Earth at all. The reader will recall that this aspect was dealt with in the previous chapter in regard to the Moon. From the Moon, however, colonists could much more easily make regular visits to Earth to prevent this peculiar form of "muscular atrophy."

The prospect of having to live forever on Mars under highly artificial conditions seems daunting. However, to persons who have known nothing else it would simply be their way of life, and if safe, comfortable, domed cities were created, then they might conceivably regard their life as quite pleasant. There would always be danger, but this would bring enhanced vigilance, thereby reducing the danger. After all, life on Earth is not particularly safe. Apart from the risk of nuclear war, people in their thousands get smashed up every year in automobiles, scores are murdered by terrorist gangs, criminals rape, rob, and kill tens of thousands, a major air disaster can easily produce several hundred fatalities, and, thanks to the way we pollute our atmosphere, oceans, lakes, and rivers, the life-span of millions is effectively shortened. Going to stay on a colonized Mars might represent a good form of life insurance.

Some of the current literature dealing with the possibilities of colonizing Mars suggests, in seeming contradiction to all the known facts, that Mars will eventually be colonized to the extent of becoming totally independent of Earth. The protagonists of this concept have even visualized the coming of a day when this totally independent Mars might war with Earth! This would seem to be an excursion into the realms of science fiction, but even if such a thing were to occur, Mars, with its domed, pressurized cities, would be peculiarly vulnerable to attack by missiles from terrestrial craft. Of course, the "Martians" could build underground cities and live a sort of troglodytic existence, but it is doubtful if even these would survive nuclear blasts.

This idea probably springs from a belief that history must repeat itself. In the past, on Earth, the colonies of major powers fought wars of independence, but a true parallel cannot be drawn here. If Mars were capable of becoming Earth's equal, some scenario along these lines might be possible. But a colonized Mars can never be anything but dependent on Earth, and that renders the idea of an interplanetary war between Earth and Mars most improbable.

Mars as a potential large-scale Earth colony does not look like a very good bet. Unlike the Moon, it does have an atmosphere, but that atmosphere is far too tenuous to be of any practical use to humans, and its oxygen content is negligible. Where did the oxygen go? Some escaped into space, and we have good reason to suspect that vast amounts became locked in the rocks of the planets as oxides and carbonates. It is still there, but unfortunately we haven't mastered the art of breathing rocks yet. The thin atmosphere also means that colonists are likely to be exposed to harmful radiation from the Sun and from space. All in all, most of the dangers found on the Moon are going to be present on Mars, with the added disadvantage that it will not be so easy to get back to the home planet. To return to Earth from a Mars then in conjunction (on the opposite side of the Sun from us) would constitute a very considerable journey indeed. To return at a good opposition still would involve a space voyage of at least 35 million miles. Mars is often described as one of the nearest heavenly bodies to us. We should really say that its orbit is one of the closest to Earth. Under unfavorable circumstances, the two planets are very far apart indeed.

The most regrettable feature of all is that of all the bodies in the Solar System, Mars is probably the best suited for terrestrial colonization. Clearly if this is the best, the worst must be unspeakable. And so it is. Long gone are the days when we could look favorably upon Mars and Venus as potential homes for Earth's surplus millions. Only out across immensity to the stars, it would seem is there a real vestige of hope.

16

Satellite Colonies

SOME READERS MAY BE wondering about the potential of some of the largest moons in the Solar System as sites for limited colonization or at least as scientific bases. Some of the satellites of the largest planets, such as Io, Europa, Ganymede, and Callisto orbiting Jupiter, and Rhea, Dione, Tethys, Enceladus, Mimas, and Titan orbiting Saturn, are themselves fairly large bodies. The first difficulty is distance. The fact that distances within the Solar System are trivial in relation to those separating us from even the nearest stars must not blind us to the fact that they are still very appreciable. Both Jupiter and Saturn along with their respective systems of moons lie much farther from Earth than does Mars. The mean distance between Mars and Saturn is over 740 million miles. They are therefore also much farther from the Sun and are thus exceedingly cold.

A brief description of each of these satellites may be pertinent in the circumstances, starting with the four major satellites of Jupiter: Io, Europa, Ganymede, and Callisto. For a long time little or nothing was known about these bodies. They were simply the four major moons of Jupiter that could be observed easily through binoculars or a small telescope.

IO

Io can be ruled out at once, for it would be difficult to imagine a more uninviting place. There is a high degree of volcanic action on its surface that is so extensive it appears to have overlaid the ancient terrain almost completely. It is still continuing. The *Voyager I* spacecraft took a long-range shot of Io for navigation purposes that revealed a curious asymmetry. The photo had been deliberately overexposed to bring out details of the background stars, and JPL engineer Linda Morabito noticed an unusual umbrella-shaped blob on the rim of the satellite. Certain imaginative scientists suggested that this was an artifact of some kind, but eventually it was realized that one of Io's many volcanoes was in the process of a particularly violent eruption when the picture was snapped. This volcanic activity may be due to tremendous stresses on the crust of the satellite brought about by the powerful gravitational pull of Jupiter. The surface of Io is very brightly colored, and this is probably due to the variety of chemical forms (known to chemists as allotropes) of the element sulphur that exist at the low surface temperatures prevailing. If Earth ever runs short of sulphur (somewhat unlikely), this would seem the ideal place to procure supplies—but it would be a long way to go!

EUROPA

Europa is also very cold and has no atmosphere. Photos have revealed a highly colored body. It appears to be covered by a number of peculiar intersecting streaks some tens of kilometers wide and extending for over a thousand kilometers or more. The cold of Europa—and, of course, this applies to all these satellites—would alone be sufficient to disqualify it for use as a colony. It might be capable of supporting a scientific base under very artificial conditions.

GANYMEDE

The surface of Ganymede bears such a remarkable resemblance to our own Moon that its geological history may have been similar. One feature it has that is not found on our Moon is a peculiar grooved terrain of parallel ridges and troughs up to 15 kilometers wide and hundreds of kilometers long. At present it is difficult to account for these. Ganymede is a desolate, airless body with incredibly low surface temperatures. It is by no stretch of the imagination a congenial spot for human beings, and we would do far better to stick to our own Moon since it is, at least, comfortingly close to Earth!

CALLISTO

The surface of Callisto is very heavily cratered with no evidence of mare-type basins, as on the Moon. Callisto also has a massive system of concentric rings. These are centered on a bright circular region 10 degrees north of the equator and lie 1,500 kilometers from the satellite's surface. Once again we are confronted by a cold, barren, airless, hostile environment, and it seems highly improbable that any humans would descend to the surface of Callisto except for a brief reconnaissance.

We cannot rule out the possibility of some form of scientific base on Jupiter's major moons—with the exception of Io. The intensely cold, stark, airless environments and the totally artificial conditions necessary for life would hardly make those assigned there feel particularly overjoyed. The possibility of rich and valuable mineral deposits on Jupiter's moons is sometimes raised. However, we do not know enough of the geology of these satellites to give a categorical answer, and minerals obtained under conditions of such immense difficulty and at such a distance would not be cheap. The science-fiction writers who have so often regaled us with tales of ore freighters playing between Earth and Ganymede (why is it nearly always Ganymede?) must not have considered the economics very seriously. It is perfectly true, of course, that deposits of certain minerals on Earth, thanks to the profligate use of them by man, are already coming into short supply, and lower-grade ores of some minerals, such as copper, are being mined increasingly. World supplies of mercury, to name one other, are considered to be sufficient only for a few more decades. But whether or not deposits of heavy metals would be found in the crust of any of these very remote satellites is uncertain. No such heavy metals have been found so far on our moon.

Next we will consider the major satellites of the lovely ringed planet Saturn. The planet has 22 satellites, and these constitute a most diverse and remarkable ensemble of bodies. However, only seven of these can be classed as major moons, and only these interest us. Six are composed largely of ice. Moving outward from the planet they are, respectively, Mimas, Enceladus, Tethys, Dione, Rhea, and Iapetus. Earth-based spectroscopy established ice as being the predominant material on their surfaces. All follow regular orbits that are almost circular and confined to the equatorial plane of Saturn. The inner five occupy adjacent orbits, but Iapetus lies much farther out, beyond the giant Titan, in a highly inclined orbit. Mimas and Enceladus are 400 and 500 kilometers, respectively, in diameter. Tethys and Dione are about 1,000 kilometers, and Rhea and Iapetus are about 1,500 kilometers.

MIMAS

Mimas, the smallest and innermost of Saturn's classically known satellites, orbits the planet in just under 24 hours. The most striking surface feature of Mimas is undoubtedly the giant crater provisionally named Arthur, after the legendary English king of that name (why King Arthur should have been chosen for this distinction is far from clear). The crater is 130 kilometers in diameter and is thought to be as much as 10 kilometers deep. A large central mountain rises 6 kilometers from the crater floor. This enormous crater is about one-third the diameter of the satellite and was almost certainly produced by impact with a body of fairly considerable dimensions—probably one about 10 kilometers in diameter. It is reckoned that the collision must have come very close to rending Mimas apart. The entire surface of this satellite displays the scars of repeated impact cratering. So thoroughly is Mimas cratered that it is impossible to increase the number of craters, since additional impacts merely destroy old craters as they create new ones. Despite the fact that Mimas is composed primarily of ice, the craters bear a marked resemblance to those on the Moon. The craters are, of course, of external (impact) origin, but Mimas also shows signs of internal activity in the shape of grooves up to about 100 kilometers long, 10 kilometers wide, and 1 to 2 kilometers in depth. However, these cracks may have resulted from the same colossal impact that produced the giant crater. Mimas has a density of 1.3 grams per cubic centimeter. Since pure ice would have a density slightly less than 1 gram per cubic centimeter, the inescapable inference is that Mimas is largely composed of ice with a small proportion of rock. Clearly no Earth colony could ever take root on a body of this nature.

ENCELADUS

Though Mimas and Enceladus constitute a pair, the two could hardly be more dissimilar. Whereas Mimas is heavily cratered and is more or less the kind of satellite to be expected in this region, Enceladus could well be described as one of the strangest and most enigmatic objects in the entire Solar System. First, Enceladus is almost fantastically bright, reflecting over 90 percent of the incident sunlight. This degree of reflectivity exceeds that of fresh snow and would indicate that the surface must be composed of ice having a very high degree of purity and devoid of contaminating rock. Enceladus is also much colder than all the other satellites of Saturn, *Voyager 2* having recorded a day-side temperature of -200°C. Parts of the satellite display impact craters up to 35 kilometers in diameter, but these are the largest craters, and there is none of the dense packing of craters manifested in Mimas. Broad swaths of the satellite show no cratering whatsoever. It is believed that mechanisms are even now at work that somehow obliterate cratered terrain. In some existing areas of older cratered terrain, some craters apparently have been altered by flow and relaxation of the crust. This suggests that the mantle beneath has retained a measure of plasticity for anything up to 20 kilometers down. In certain places, the crater-free areas show undulating ridges that could conceivably be flow marks resulting from long past outflows of water. In view of the low temperatures prevailing, it seems reasonable to assume that this water must have been heated internally. This could be due to volcanic or magmatic action. One mountain that could be of volcanic origin has been observed. Enceladus will never be a prospect for colonization, but in view of its unique geological features, it may one day have temporary or occasional bases from which geologists will make a detailed and intimate study of the satellite's physical aspects. The conditions under which the personnel of such a base would have to exist would be rigorous and unpleasant in the extreme.

TETHYS

Tethys resembles Mimas, for Tethys' surface shows clear indications of heavy impact cratering. One crater has a diameter of 400 kilometers, which is equal to the entire diameter of Mimas. The most interesting feature on Tethys is an immense complex of valleys, known as Ithaca Chasma, that extends three-quarters of the way around the satellite. The width of this complex is about 100 kilometers, and its depth is several kilometers. It is probable that Ithaca Chasma is a system of cracks produced by the impact that was responsible for the 400-kilometer crater. Alternatively it could be due to expansion of the satellite as its liquid water interior cooled and then expanded as it froze. No terrestrial colony is remotely possible.

DIONE

Dione has the distinction of possessing the highest density of all Saturn's icy satellites (1.4 grams per cubic centimeter), thus indicating a higher proportion of rocky materials. One hemisphere is heavily cratered; the other appears to be covered by a system of valleys. The satellite is totally inimical to any form of human colonization.

RHEA

With a diameter of 1,500 kilometers, Rhea is the largest of the inner satellites of Saturn, though it shows less evidence of geological activity than the others. Like Dione, the "leading" hemisphere (as a celestial body orbits another there must be a "leading" hemisphere and a "trailing" hemisphere), being more protected, shows streaks that probably represent valleys. The "trailing" or cratered hemisphere has been likened to the highlands of the Moon, though in this instance the material is brilliant white ice and not dark brownish rock.

IAPETUS

Iapetus is the outermost of the large icy satellites of Saturn. It is virtually the twin of Rhea being 1,450 kilometers in diameter, though it differs significantly in density, suggesting a smaller proportion of rocky constituents. It has a dark "leading" hemisphere and a bright "trailing" one. Most of the surface is believed to consist of ice, though the "leading" face (except near the poles) is as black as tar. This is probably due to a coating of very dark material on top of the ice. When an occasional meteor strikes this dark face it would be reasonable to expect white spots (exposed ice) or even bright ray craters to result. In the absence of these we must assume that the dark material is either thick or is being continually renewed and thus obliterates bright impact ejecta. Whatever the explanation, it is clear that a cold, stark, airless body such as Iapetus could never constitute a colonizable sphere.

TITAN

The really interesting satellite of Saturn is inevitably the giant Titan, which has been described as the "Earth-like satellite," though this description should not be taken too literally. It is a truly outstanding body and, with a diameter of 5,800 kilometers, is intermediate in size between Mercury and Mars. Thus, though technically a satellite, it is one of planetary dimensions. Moreover, it is exceptional in that it possesses an atmosphere, but not one that human beings could breathe. Nor is it a thin, tenuous, negligible atmosphere for, in respect of mass and surface pressure, it exceeds that of Earth. The atmosphere is opaque and is believed to consist of multiple layers of aerosols ranging from tenuous, smokelike hazes to deep clouds of either liquid or frozen methane. The surface temperature is believed to be about -180°C. Due to the opaque atmosphere, all other data concerning the surface and interior of Titan must be determined indirectly. Its density is about 1.9 grams per cubic centimeter, and this implies 45 percent ice and 55 percent rocky and metallic materials. In its size and structure Titan is similar to the Jovian satellites Ganymede and Callisto. Nevertheless, some factor in the formation of the satellites must have been very different. Why else should Ganymede and Callisto be totally devoid of atmosphere while Titan has a thick one of methane and nitrogen? A possible explanation is that Titan was formed in the colder parts of the solar nebula and managed to incorporate "ices" of ammonia and methane as well as ordinary ice. As the temperature of Titan's interior rose due to coalescing of matter and the presence of radioactive materials, these gases escaped and produced a primary atmosphere. The effect of sunlight on ammonia (NH_3) resulted in its reduction to nitrogen and the loss to space of the hydrogen, while the more stable methane (CH_4) survived to the present time. It is well known that methane and nitrogen can react under certain conditions to form complex organic compounds, and this probably occurred in Titan's atmosphere and on its surface.

Infrared spectroscopic examination of Titan's atmosphere has indicated the presence of methane (CH_4), ethane (C_2H_6), acetylene (C_2H_2), ethylene (C_2H_4), methylacetylene (C_3H_4), propane (C_3H_8), and diacetylene (C_4H_2). All these hydrocarbon molecules can be produced by the action of sunlight on methane. Despite its considerable distance from the Sun, sufficient solar radiation to render these reactions possible must have reached Titan. Certain nitrocarbons also have been detected. Among them are hydrogen cyanide (HCN) (this is the deadly poison known as prussic acid and gives us good reason for not immigrating to Titan), cyanoacetylene, (C_3HN), and cyanogen (C_2N_2). Titan therefore provides all the necessary building blocks of complex organic compounds. Could these one day lead to the production of living cells? It is a fascinating thought.

About 30 or 40 years ago it was not uncommon for astronomical artists to portray Titan as a rock- and icebound world under a clear, deep blue sky (indicating presence of a transparent atmosphere) in which hung, in all its glory, the magnificent ringed Saturn. Sadly this is a conception that must be consigned to the astronomical trash bin together with the Martians and canals of Mars.

Because of Titan's thick and opaque atmosphere we can only speculate on its prevailing surface conditions. In the light of existing evidence we might envisage a situation roughly analogous to that of Earth, with methane in the role of water. There would be oceans of liquid methane and polar caps of solid methane. Rain would be falling from the upper atmosphere but it would be composed not of water but of organic compounds that may have produced a 100-meter-thick layer of tarlike materials. Little of the weak sunlight impinging on the planet's upper atmosphere will reach the surface, and it must therefore be a very gloomy and forbidding place. Despite these features, Titan is considered by many astronomers to be more like Earth than any other planetary body in the Solar System, Mars not excepted. Titan has been described as "a planet in deep freeze." Titan is almost certainly the most interesting body in the Solar System in which to study the processes that eventually gave rise to life on Earth some 4,000 million years ago. It is not, however, a place to which humans could ever immigrate, and even brief scientific expeditions to the surface of Titan, though they would be of immense value and interest, will forever be fraught with deadly peril.

So there we have it. It is no use looking further around the Solar System for places to immigrate to. They do not exist. When man wishes to extend his domain, he must head for the stars. There is no real alternative. It is doubtful if there ever can be.

17

Secondary Colonization

IT MIGHT BE OF INTEREST to consider secondary colonization, by which we mean the establishment of fresh colonies by the descendants of original colonists. This is looking very far into the future. The difficulties involved in establishing an original terrestrial colony several light-years from the Solar System are very real, and clearly it will be a considerable time before that colony could develop a technological base that would allow it to repeat the performance. And why, the reader is surely entitled to ask, should any of its members be so inclined? The positive form of this question might be couched in the following terms: "We, or at least our ancestors, succeeded in crossing the gulf of space to another star. Why should some of us not emulate that feat? Is it not now time to extend terrestrial civilization farther into the galaxy?" Here is the rationale for what has been called "creeping colonization." However, unless rapid systems of interstellar travel through hyperspace (that is, via other dimensions) have been perfected (assuming this is feasible), it is most unlikely that terrestrial colonization is going to pervade the entire galaxy. And, on reflection, that might be just as well, for the chances of blundering into a supercivilization of aliens would be vastly increased by such an expansion. Star wars provide first-rate entertainment in the movies. The reality, we suspect, would be greatly and horribly different.

Man might have quit Earth and the Solar System to escape persecution or because Earth was becoming grossly overpopulated or unacceptably polluted. Much as we might hope that with the passing centuries men might learn wisdom, these afflictions could occur again on the colonized world. This explanation for further colonization is not strictly necessary, however. It seems quite probable that *Homo sapiens* would once again wish to spread his interstellar wings simply because the stars, other stars, were there and beckoning. Thus a terrestrial colony having proved successful might become the progenitor of new Earth colonies at distances from the Solar System greater than that of the original leap. This could represent a multiple process, for on this occasion two or three colonizing expeditions might set out, each bound for a different star. It might also be an ongoing process, a sort of "star-hopping" by which terrestrial civilization is spread among the stars of our local star cluster within the Milky Way. At each step, of course, the difficulties would increase. We must set these developments in their proper time scale. Some have suggested that this would be as long as the geological time scale. There is little doubt that this is much too long. Certainly we cannot imagine, much less foretell, what the course of evolution will have done to us over a period of such magnitude. We cannot be certain we will not be extinct. But, if we still exist, it seems likely that we will have adventured into the galaxy as far as our technological capacities will take us.

Even if secondary colonies are successfully established, the risks invoked in such colonization would almost certainly have been greater than those to which the original colony was subjected. An original colony would have had the resources of Earth upon which to draw, though, as we have seen, these would nevertheless be limited by the amount of material that could be carried by the star ships. A secondary colonizing expedition would, on the other hand, be leaving from a planet the resources of which might not have been developed to a comparable extent, even though the society there felt itself able and ready for another interstellar leap. As before, once beyond the point of no return, it would be completely on its own. Much has been made in the pages of science fiction of a colony being able to draw upon Earth for continued supplies and succor. This seems totally unrealistic unless some "rapid transit" method of interstellar travel can be established. Should such a thing become feasible, the situation would be changed out of all recognition, with consequences that are impossible to predict. For now, such systems of interstellar travel remain an intriguing and tantalizing dream. For all we know, other highly advanced technological civilizations within the galaxy may already have discovered the secret. If so, the reader may be tempted to ask why haven't we had a visit. It is a good question. All we can say is that the galaxy is a big place. They may just not have stumbled upon us—yet!

If terrestrial colonization of our local star cluster eventually occurs, the spread of our civilization will, of necessity, prove very irregular. We cannot see it taking place in the manner in which a stone flung into a pool creates ever-widening concentric circles. Stars that are conveniently "near" (the term is relative) will not all be like the Sun. We know that many are not. The nearest of all stars, Proxima Centauri, is a class M "cool" red dwarf, and even if it should possess planets (which is highly unlikely), it is even more unlikely that they would prove colonizable. The twin system of Alpha Centauri (of which Proxima Centauri may be the third component) is also doubtful, largely because it is a binary (twin sun) system. One of the components is a type G star with a surface temperature of 5,750°K and therefore closely akin to the Sun, though with a marginally larger diameter. The companion star is of type K with a surface temperature of 5,000°K and a diameter of 750,000 miles. Though not a solar-type star it is not too far removed from this category and could be acceptable. Unfortunately, the orbits that planets within binary systems would pursue might prove highly eccentric—colonists could be alternately frozen and roasted! Much would depend on the degree of separation between the stars in such systems. If the separation were large, planets could orbit each individual star much as the planets of the Solar System orbit the Sun. But if the individual components lay relatively close to one another, the effects would probably render the colonization of planets impossible.

If from a vantage point in space thousands of years hence we observed the pattern and spread of terrestrial colonization, how would it all look? The answer probably is "random," with "tentacles" and "subtentacles" leading in several directions, with the

Sun roughly in the center. In some directions these "tentacles" would be longer than others simply because more appropriate stars and planetary systems happened to lie in these directions.

Somewhere and at some time a terrestrial colony could fail, which, under the circumstances, presumably means it will suffer extinction. This might be due to disease, to some cataclysmic physical disaster, to lack of resources, or to prolonged imbalance in the sex ratio. If humans continue to expand outward, the law of averages will produce a failed colony sooner or later. Whether evacuation of a colony that was visibly failing could be achieved is highly debatable. Could a "neighboring" colony come to its aid? "Neighboring" might imply a distance of 5 light-years and, on the interstellar scale, that is close. It also represents a slight matter of 30 million million miles. That is a long way for a distress signal to carry and a long way for a rescue mission to come. The fact is that there might be no rescue from another, more successful colony. This is one of the harsh realities of interstellar colonization. It is virtually impossible to find a way around it unless really rapid transit via curved space or another dimension becomes feasible. The "galactic empires" so beloved of science-fiction buffs are thus undone by the realities of space and time.

Each planetary colony is going to be a separate, self-contained entity. Radio signals might pass between them, but messages relayed by this medium can travel only at the speed of light. Consequently, years must pass before such a signal reaches its destination—and an equal time would be required for a reply.

It has been suggested that a number of relay stations might be placed in the interstellar abyss between two "neighboring" planetary colonies. However, when we begin to examine the practicalities and implications, we realize that this in itself represents a positively gargantuan project. If two colonies are located on the planets of stars 5 light-years apart (and remember, this is a short distance by interstellar standards), just how many of these relay stations would be required? If we suggest four at one light-year distance apart, this is hardly enough. A light-year is roughly 6 million million miles. Just to reach the first relay station would represent a considerable achievement in itself. If we decide that by then technology has enabled our descendants to move at velocities that are high submultiples of that of light,

Einsteinian time-dilitation would begin to apply. (The strange implications of that theory will be discussed in Appendix III.) It seems we must reject the concept of relay stations. Interstellar colonization can be achieved, but it will almost certainly have to be done by cryogenic "deep freeze" or by "generation" means—unless, of course, a way can be found to travel quickly through another dimension. Earth never could rule over a galactic empire, but it might conceivably place a number of "other Earths" within a strictly limited distance of the Sun. Until we can find some kind of interstellar "seven league" boots, this must suffice. Some may consider this pessimistic, but I suspect "realistic" is the more appropriate word.

There is one rather intriguing development that may someday occur. Let us suppose a colony did become firmly established on a near replica of Earth 5 or so light-years remote. Over several centuries it will have developed a sturdy technological and industrial base and perhaps it will show sufficient wisdom not to waste its time, life, and resources in stupid, fratricidal strife. To its members Earth would be only a name, the details of which would be found only in the data banks of the colony's archives. Suppose that some of the inhabitants of the new world were seized with longing to see the planet of their origin. An expedition would set forth, again by cryogenic means, to visit Earth. Just how would its members be received, or what would they find? Those are difficult questions. Would they come upon a nuclear wasteland, the terrible result of the very holocaust their ancestors had quit Earth to avoid? And if this supreme tragedy had been averted, would they be welcomed by the people of Earth—or spurned? Here we pass into the realm of speculation.

18

Colonies in Space

HAVING LOOKED BRIEFLY at the development of a terrestrial colony under the highly artificial conditions that would be necessary in a lunar, Martian, or satellite environment, it seems appropriate to close this book by dwelling at some length on an entirely different and even more artificial form of terrestrial colonization.

One scheme long regarded as a future possibility to span interstellar space is "generation travel." A vast IST would set out for a "close" star such as Alpha Centauri (4.3 light-years distant) on a journey scheduled to last about 200 years. Obviously the first generation of space travelers will not live to see the planet of the star that is the mission's destination. Neither, for that matter, will the second, third, fourth, and fifth. The eighth generation will reach the chosen planet when they are in their prime and the sixth and seventh generations will see it before they die. The second to fifth generations will know only the huge IST as their home. To all intents and purposes, the star ship will fill the role of "planet" for them, albeit an odd one by accepted standards. They will, in effect, be colonists within an artificial body. It is obvious that this will create many severe problems. Some of these I dealt with in *Journey to Alpha Centauri* (1965). There were problems I omitted at that time, however, and I will discuss them here as well as consider the general necessities of generation travel.

To all the midgenerations the great IST would be much more than just a traveling colony in space. By necessity it would constitute a closed ecosystem—that is, a life system in which nothing can be replenished and in which all waste must be recycled—a system, in other words, to which nothing can be added and from which nothing must be removed. This is not the most pleasant of subjects, for it must deal not only with human requirements but with such items as human waste. If we are to have a closed ecosystem, waste of all kinds—human, animal, and vegetable—cannot just be dumped in space. If it were, the system could not be defined as closed. Besides, in a closed ecosystem such waste represents valuable raw material. Let us dwell on a number of interrelated factors, including human population, animal population, plant production, waste processing control, and industrial products.

The provision of food in space is likely to evolve through four fairly distinct phases that for convenience we will refer to as A, B, C, and D. Phase A refers to food that is not produced in space but is taken aboard the IST prior to departure from Earth. During this phase human waste is either jettisoned overboard into space or returned to Earth for conventional disposal. In Phase B most of the food used is still that loaded prior to departure, though human diet is probably by now being supplemented by plants grown hydroponically during the early stages of the journey. By the time Phase C is reached, the closed ecosystem is really established. Food consumed is that now being grown exclusively within the ship. Water and gases are being recycled. A slight loss of gas is almost inevitable, and thus a certain degree of replenishment from cylinders on board the ship will be necessary. This applies particularly to the gas nitrogen. In the circumstances the ecosystem cannot wholly be regarded as closed. It is doubtful if this could ever be achieved, but it is near enough to merit the title. In Phase D food is now being produced in large quantity by a hydroponic system. From Phase C onward it will be apparent that all reliance on Earth has ceased.

On a "generation" IST the population would comprise young people (about 20 to 30 years old) of varying occupations and professions, with both sexes fully represented (essential if reproduction is to continue at a satisfactory rate). The middle generations on a star ship to Alpha Centauri will simply be life members of a permanent colony in space. In overall terms, of course, such a colony is nonpermanent since one day, with reasonable fortune, it will reach the destination world, and those aboard will disembark.

Those humans for whom the ship is their only world will have nutritional and other requirements. We should consider the daily input and output of humans. Input covers dietary energy requirements, protein, oxygen, and water requirements. Output comprises solid and liquid wastes, nitrogen (in chemically combined form in liquid waste), and carbon dioxide (exhaled). As for water requirements, it is estimated that only 5 to 6 percent is required for drinking and cooking, the remainder being used for washing and other purposes.

It is necessary to consider the total volume of the star ship in relation to atmospheric requirements. The atmosphere must, of course, be a mixture of oxygen, nitrogen, and carbon dioxide equal or closely akin to the atmosphere of Earth at sea level. Total change of atmosphere is necessary once every 12 hours to maintain the gas mixture in the required proportions. The human metabolism generates about 0.5 kilogram of water per person per day, and this has to be removed from the atmosphere either to supplement water supplies or for transfer to a waste processing unit. In space, waste is so valuable that it should not really be termed waste.

A closed ecosystem in space will also require that the corpses of deceased colonists be treated in a less reverent manner than would be the case on Earth. In *Journey to Alpha Centauri* I conveniently "forgot" this unpleasant truth and merely had the bodies placed in special pressureproof containers and launched into space to drift forever under the eternal stars. Very poetic but not at all sensible. In an artificial planet or "generation" star ship the bodies' constituents on death must be retained if biomass is to remain constant. Cremation, already gaining increasing popularity on Earth, would be acceptable. The incinerator of the waste processing unit would permit recovery of all constituents. The religious aspect must also be considered. Christian religions teach us that after death the spirit of the dead person goes to heaven, purgatory, or hell. The body will be raised from the dead at the Last Judgment, but until that time we cannot regard it as strictly relevant whether it disintegrates slowly in the soil or quickly by cremation. Most Christian churches have taken a more liberal attitude toward cremation in our time. Other religious bodies would have in some cases more, in some cases less, difficulty adapting to this practice.

So much for biomass as regards the human population. We must also consider plant production and its ramifications. On Earth, under the rays of the Sun, photosynthesis takes place. This is the basis of food production and a key process in element recycling. In a "generation" star ship colony there is, of course, no sunlight, so photosynthesis, if it is to take place, must find an alternative source of "sunlight." There do exist nonphotosynthetic methods of plant production, but these merely change one form of organic matter into other forms having a higher nutritive value. Only photosynthesis can create organic food from inorganic materials. Photosynthesis can be defined as the synthesis of simple carbohydrates from carbon dioxide and water, with the liberation of oxygen, using the energy of light, in green plants, chlorophyll being the energy transformer. In the type of "enclosed" colony we are presently considering it is easy to see the value of green plants, since they absorb human-exhaled carbon dioxide when they are exposed to sunlight, at the same time releasing valuable oxygen. On Earth during darkness the opposite effect takes place, carbon dioxide being produced and oxygen being consumed (which is why it is not a good idea to have your bedroom festooned with plants and flowers). On the star ship it would be easy to arrange for growing plants to be exposed to light similar in spectrum coverage to sunlight all or most of the time.

It has been suggested that crops could be grown using lunar regolith as soil. Under space conditions a form of hydroponics probably would prove more convenient. Regolith is attractive in that it allows normal rooting and physical support of plants. Unfortunately, roots and other organic matter would, over a period of a few years, tend to accumulate, thus acting as a large-scale though temporary "sink" for many elements that should normally be helping to feed and maintain the plants.

We discussed hydroponics in Chapter 10, though the method described there represented a simple, basic technique for production of plants on a relatively small scale. To obtain a large crop a much more sophisticated form is needed. One system, already in commercial use on Earth, is used for the production of vegetables. It is known as the nutrient film technique. Plants are supported, but in such a manner that their roots lie in shallow gullies through which the nutrient solution flows. This nutrient solution contains every element that the plants require, and in the correct proportions. The method is advantageous in that it allows the composition of the solution to be monitored and adjusted as necessary. In the main, the solution consists of nitrogen, phosphorus, potassium, calcium, magnesium, sulphur, sodium and chlorine with iron, copper, zinc, boron, and molybdenum as essential "trace" elements.

To speak of harvests being produced within the metal shell of an IST seems mildly ridiculous. The term "harvest" inevitably conjures up an image of great, wide, open spaces. The size of the harvests produced aboard "generation" ISTs would not, of course, be comparable. Nevertheless, they would have to provide for the midgenerations of space voyagers who would be born on the vessel, live their lives on it, and eventually die on it. Such ISTs would surely have to be vast. Here is an instance in which science fiction may become science fact. The incredibly vast star ships we presently see on TV and movie screens must be regarded as precisely what would be required. "Cryogenic" ISTs, with their occupants in a state of suspended animation, might conceivably be smaller since no crops need be grown in them. However, immense quantities of essential supplies would still need to be carried, and as Chapter 10 has demonstrated, two or three ships equipped to produce food hydroponically would be of inestimable value. So perhaps "cryogenic" ISTs would be nearly as large. Such gargantuan craft could be constructed only in space; hence the peculiar nonstreamlined design with which, thanks to science fiction, we are already familiar. On arrival at their respective destinations such craft could not be brought down through the atmosphere to a planet's surface. They would remain permanently in a (hopefully) nondecaying orbit.

Since crops vary very widely in their energy content as well as in the protein, minerals, and vitamins they contain, no single type of crop could be expected to satisfy all nutritional requirements or to provide a satisfying, balanced diet. Choice of crop species would be dictated by a number of factors, the chief of which would be:

- potential crop yield
- time taken from sowing to harvest
- nutritional value of crop as related to human nutritional requirements
- human preferences
- quality of waste plant material that could be fed to animals (thus determining the contribution of animal products to human diet)

It is possible to determine a mix of crops that satisfy human nutritional requirements while keeping total crop area to a minimum.

Animal feedstuffs come in diverse forms and include plant roots and stems, algae, and waste animal products (parts of animals that humans usually don't eat). Such a range in types of feed would suit a variety of digestive systems including all ruminant species (for example, cows). Omnivores (those capable of digesting both plant and animal material) could consume plant material as well as waste animal products.

It is not improbable that the concept of slaughterhouses on IST's would be highly repugnant to people of the 21st and 22nd centuries. In their absence the colonists would, of course, rely almost entirely on a vegetable diet. Animal products such as milk, cheese, and eggs would also be available.

Two animal species seem well adapted to these requirements: cattle and chicken. Friesian dairy cattle would be ideal because of their highly efficient conversion of food protein, their ability to utilize low-grade plant material, and their high production of milk. Pigs would be of interest because of their ability to digest both plant and animal wastes, but since they provide neither milk nor eggs, their fate would be very clear. The colonists might wish to bring them anyway, just as they might wish to bring small numbers of other domestic animals such as dogs, cats, and horses, which would be of little use on the journey to Alpha

Centauri but would be highly desirable once they settled their new world.

Dead animals in a closed ecosystem must, like dead humans, be replaced. Thus the animals will need to reproduce their replacements.

In a closed ecosystem there would need to be a number of highly efficient waste processing units to process a wide range of materials into usable forms. Processing procedures would consist most likely of a number of sequential stages, including drying, crushing, biological oxidation, incineration, and gas "scrubbing" (purification).

We must now put aside squeamishness and turn our attention quite directly to the question of human waste. The average production of urine per person per day is 3.25 kilograms. Since it contains many elements in inorganic form, it is an ideal base for the nutrient solution used in the production of plants hydroponically. The composition of human, cattle, and pig urine tends to vary, largely due to differences in diets. The nitrogen content of human urine is put at just over 19 grams per person per day, whereas that of cattle and pigs is lower, at 9 and 8 grams, respectively per day. Urine composition also varies on a day-to-day basis. The other valuable elements in urine are phosphorus, potassium, magnesium, and calcium.

Much of what has been said regarding liquid waste is true also of solid waste. A discourse on this subject is even less pleasant, so we will not offend any reader's sensibilities by dwelling at length on it. It is, of course, too valuable to waste. Even on Earth, animal manure returns valuable constituents to the natural cycle.

Plant waste dry matter and algal dry matter are extremely useful as animal feed. They would first need to be pulped and pressed so as to effect the partial removal of protein. Up to 75 percent of plant protein could be extracted. This would be extremely useful as dietary protein for young cattle and pigs, the residue being used as ordinary cattle feed.

Over half the total of animal waste would, as might be expected, be in the form of bone. It is first necessary to separate bone from other organic material such as scraps of fat. Ground bone is a very rich source of the elements calcium, magnesium, and phosphorus for cattle and sheep. Some ground bone could also be used to provide these elements in the plant nutrient solution mentioned earlier. Material other than bone could be partially dried prior to its being intimately mixed with plant waste. After milling this could be pelletized, a form useful for feeding purposes.

All carbonaceous waste that could not be disposed of in other ways would be incinerated. These would include anaerobic digestion residues and some industrial waste. As mentioned earlier, the gas methane produced during digestion of sludge could be used as fuel for the incinerators. Incineration, of course, results in the production of gaseous products that must be passed to a gas scrubber for purification. The solids remaining after incineration could then be treated to provide water-soluble forms of calcium, phosphorus, and other elements for use in plant nutrient solutions. At the same time, such undesirable elements as lead and cadmium could be removed.

It is now time to think a little more about the atmosphere within the IST, especially as regards carbon dioxide content. Humans and animals produce this gas through the normal process of respiration. In a different manner so would the various waste processing units. Plants, on the other hand, consume carbon dioxide during photosynthesis. It has been estimated that humans and animals between them would, due to exhalation, produce 2.56 kilograms of carbon dioxide per person per day. At the same time, waste processing units would produce the equivalent of 0.35 kilogram per person per day. Thus input of carbon dioxide into the atmosphere is equal to 2.91 kilograms per person per day. Plants, it has been estimated, would consume the equivalent of 2.90 kilograms per person per day of carbon dioxide. A balance is thus obtained. However, we must also consider the question of carbon dioxide and oxygen circulation. It seems highly unlikely that "generation" space travelers would wish to share their living quarters with livestock. (The livestock probably would not mind; the humans assuredly would!) It also is hardly likely that humans would be too anxious to live cheek by jowl with plant production or waste processing units. In other words, human, animal, plant, and waste units would be totally and hermetically sealed from each other. This being so, carbon dioxide concentrations in the human, animal, and waste processing units are going to build up, whereas in the plant production section there will be less and less carbon dioxide available for the plants and increasing amounts of desirable oxygen. To obviate this a simple form of atmospheric circulation has been suggested whereby gas from the human and animal quarters would be passed to the waste processing units via the gas scrubbers to remove any pathogenic agents before being filtered and transferred to the plant production sections. In turn gas would be pumped from the plant production sections to both the human and animal sections. The rate of atmospheric circulation between the sections would be controlled to permit complete mixing, with a constant carbon dioxide concentration of approximately 0.03 percent by volume.

There is a very great deal that could be said on the theme of closed ecosystems in space colonies, and in this chapter we have done little more than scratch the surface of a complex subject. However, what has been said should enable the reader to grasp the fundamentals and to appreciate at least some of the far-reaching ramifications. To those who might like to explore the subject in much greater depth there are a number of very good papers on the theme that go deeply into the chemical, physical, and biological processes and that quantify them. The one most recommended is "A Closed Ecosystem for Space Colonies" by I. R. Richards in *Journal of the British Interplanetary Society,* Vol. 34, (September 1981), pp. 393–99. This is an excellent piece of constructive thinking to which its author has devoted much time and talent.

Epilogue

IT'S A REASONABLE ASSUMPTION that, in so large a subject as the colonization of other worlds, certain aspects may not have been covered as well as they deserve. Also, a few readers may not wholly agree with some of the conclusions reached. This is understandable, and I fully respect the alternative conclusions reached by others. Merely to consider planetary colonization has an effect upon the way we view every facet of human endeavor, past and present.

Many readers may wonder why an Earth-like planet has been envisaged in most of the chapters of this book, but the reason is perfectly straightforward. Only on a planet that is a reasonable facsimile of Earth could large-scale, permanent colonization be accomplished. Humans could exist only on the Moon or Mars in small numbers and under the most highly artificial conditions. Such "colonies" of the perhaps not too distant future must really be seen as scientific research establishments in which the personnel would be changed at fairly frequent intervals. Science-fiction writer Arthur C. Clarke, in his gripping and excellent novel *A Fall of Moon Dust,* portrayed the Moon, sometime during the next century, as a place where some of our descendants could spend brief and fascinating vacations and where some people might have been born and become permanent residents. But never at any point in the story did he suggest that conditions for life were anything but artificial in the extreme—a place where the price of survival was eternal and unremitting vigilance. The

Moon is very close, astronomically speaking, and it is by no means impossible that Clarke's vision might eventually be realized. On the other hand, Mars, 35 million miles distant at its nearest to Earth, is a totally different proposition. To go yet farther afield, a scheme that we must suppose is theoretically possible has been suggested by Dr. Carl Sagan for the terrestrialization of Venus. This was dealt with in my *Where Will We Go When the Sun Dies?* (Stein and Day, 1983). Clearly the high temperatures on Venus would be a considerable handicap to any attempts to colonize the planet.

Some may wonder why we have assumed that only a relatively small colonizing mission consisting of a few ships will be sent. The assumption was made to emphasize the difficulties and problems that would confront terrestrial colonists no matter how Earth-like the planet. A truly colossal mission sponsored and supported by, for example, all the nations of Earth under the aegis of the United Nations (assuming that body ever became sufficiently "united") with scores and scores of ISTs able to transport from Earth vast and varied mountains of equipment and supplies would certainly minimize the difficulties and complexities of creating a terrestrial civilization on a new Earth. It is possible too that, for a variety of reasons, a proportion of such a great armada could be lost en route. Who can say what unknown perils lie in the depths of interstellar space? Of the host that left Earth, only a handful might reach the chosen planet. Naturally, a small mission like the type we have been envisaging also could suffer losses. In that event none of the ships might arrive. There would be no colony. That we accept. No great adventure or project is without risks—and surely colonization of a planet of another sun several light-years distant would constitute the greatest project ever conceived and attempted by man. We must assume the risks would be commensurate with the magnitude of the project.

The reasons given for quitting Earth and embarking on a mission embodying so many imponderables might also be open to question. In *Where Will We go When the Sun Dies* such a project was seen as mandatory and, assuredly, if some cosmic disaster were about to overtake our life-giving luminary the choice would be simple: Go or perish! Fortunately, it seems that for at least the next few billion years, this is the least likely possibility.

Journeying to the stars just because they're there would reflect mankind's love of a challenge. An interstellar journey taken on that basis probably would be carried out by a small group of dedicated men and women willing to sacrifice their lives in the interests of science.

The desire to escape an overpopulated and overpolluted earth is a very sound and reasonable reason that is all too likely to be given within the next century or two as a justification for undertaking interstellar colonization.

Finally, tyranny on Earth and within the Solar System seems as likely a reason as any for a relatively small group to stake their all on the chance of establishing liberty and justice on a new, unspoiled world rather than continuing to endure a brainwashed existence on Earth. Could such a tyranny ever envelop the Earth? A mere 40 years ago one tried and nearly succeeded and at the present time another, of differing ideology, seems intent on the same goal. But for the strength, resolve, and purpose shown by the United States and its principal allies, that tyranny might already have prevailed. If the West found itself unable to contain such a creed, then a terrible darkness would envelop the world we know. Better by far to head for the stars—or die trying.

Appendix I: Cryogenic Hibernation

WE CAME TO THE CONCLUSION in an early chapter that colonists from Earth to a terrestrial-type planet orbiting another star would most likely invoke the method of interstellar travel known as cryogenic hibernation or more popularly as the "Big Sleep." It is worthwhile discussing this technique in detail.

The term "Big Sleep" finds considerable usage, but it is important to explain that the use of the word "sleep" is erroneous, since hibernation is an entirely different process. During true sleep body temperature remains normal, with bodily functions proceeding at their normal rate. Hibernation is quite different, one important manifestation of this being the fact that "awakening" from hibernation to a state of total awareness is much more prolonged than waking from normal sleep.

At the present time, medical science is unable to prolong the duration of the human life-span to any significant extent. The eradication of certain diseases has prolonged the average life-span, but that average is still just a little over the biblical three score years and ten, and a human being who reaches his hundredth birthday is still a cause of astonishment. Besides, virtually all people who reach that age have mental and physical powers that are markedly degenerated. The technique of suspended animation, if perfected, would permit a human being to prolong his life-span almost indefinitely. Of course, such a life-style would be a little odd. One could, for instance, spend protracted periods in the suspended state, waking for a few weeks or months at preselected intervals. By this process a person could distribute his life over centuries.

There are two possible ways, according to present knowledge, in which we might approach the problem of suspended animation: by cryogenics or by hibernation. It is unfortunate that the generally accepted term "cryogenic hibernation" renders the two synonymous.

CRYOGENICS

Cryogenics (in this case we should term it cryobiology) is concerned with biological activity at abnormally low temperatures and with the effect these low temperatures have on animals and plants. These would be temperatures below the freezing point of water (0°C). Since freezing ensures the total cessation of body functions, a space voyager would theoretically require no other sustenance than the maintenance of this low temperature. A star ship would be a great, traveling, refrigerated container comprising a host of individual sealed cells or cubicles, each one containing a space voyager in a state of deep freeze. There would, of course, also be equipment for maintaining the occupants in this state, and that equipment would have to function continuously and flawlessly for a couple of centuries or more. The occupants would be thawed out as they approached their destination world and hopefully they would be in first-rate condition.

At the present time the freezing of human beings has not been perfected, and experiments along these lines on animals have not given cause for much optimism. However, we are talking of a journey that would occur fairly far in the future. The early rocket experiments of American pioneer Robert Goddard were disappointing, but nonetheless his highly sophisticated descendants in that field eventually placed men on the Moon and took unmanned vehicles to Mars. Experiments on hamsters a few years ago revealed that none survived more than a few hours after defreezing if the temperature of the fluid in which they were immersed dropped below -10°C (the temperature of a cold night in winter). None recovered when their deep body temperature fell much below -1°C (which is not very cold). None recovered fully if more than 50 percent of their body water had turned to ice or if the freezing period exceeded an hour.

The nature of the problem centers on the damage resulting not only from freezing but also from thawing. Different portions of the body cool at different rates, and so far it has not proved practicable to ensure that all parts of the body attain the same temperature at the same time. The formation of ice crystals disrupts both cells and body organs. This results in a harmful concentration of salts, and it could also harmfully effect the digestive juices of deep space voyagers. The overall consequence for a defrozen human being might be chronic stomach ulcers due to the corrosive effects of the hydrochloric acid in his or her digestive system.

Damage to the brain probably represents the most serious problem of all. Whereas body cells in general tend to regenerate themselves in the course of a year, those in the brain do not. Damage in this area is irreversible. The number of brain cells (neurons) is determined during childhood. The number then begins to diminish at an approximate rate of ten thousand per day. Though not all neurons are employed at any one time, the total number not only diminishes with age but also damaged cells cannot be replaced. Clearly it is imperative that brain damage be averted for, failing that, a thawed-out space voyager, though physically fit in all other respects, would be reduced to a virtual vegetable totally incapable of coherent thought. The brain is an organ easily damaged, and if the circulation of blood to it fails for as brief a period as 30 seconds, irreversible damage can result. Admittedly the position is marginally alleviated at lower temperatures, and doctors now estimate that at 10°C below normal body temperature, circulation may be halted for about 8 minutes without resultant damage. But even 8 minutes will hardly suffice for a journey to Alpha Centauri or any other star.

It must be admitted that we are still a long way from successfully freezing human beings. No doubt they could be frozen now, but apart from any other problem they would suffer embrittlement and would be liable to shatter if dropped! It is difficult to say what developments the next century or two may bring in cryobiology. Despite the present pessimistic outlook it is only fair to state there are instances in which human beings have experienced considerable reduction of body temperature and have survived to tell the tale. In 1957 a girl lay outdoors in an air temperature below -20°C. Despite severe frostbite, which is hardly surprising, she survived. It must be added however, that, prior to this, the young lady in question had been treating herself rather liberally to the contents of a bottle of Scotch. This indeed was the reason for her unpleasant nocturnal adventure but the presence of alcohol in quantity in her blood stream may have had a distinct bearing on her survival. A crew member from the giant liner *Titanic* in 1912 survived for a quite considerable time in water so cold that it killed most others within a matter of minutes. But, prior to stepping off the stern of the great ship as it took its final plunge, he had consumed several bottles of gin! He too lived to tell the tale. We are not suggesting however that our young men and women planetary colonists of the future should first become thoroughly inebriated! It is obvious though that "freezing" presents several very formidable problems.

HIBERNATION

Hibernation is a fairly familiar process in the animal world, and a number of creatures indigenous to temperate latitudes go into this peculiar sleeplike trance with the advent of winter, becoming conscious again only when the first warmth of the spring sun is felt. This process might commend itself to *Homo sapiens*. There seems much to be said for a state that would render us oblivious to the rain, wind, snow, sleet, hail, and frost of winter! Regrettably it lies outside our power at present. A creature in hibernation, though it may seem to be asleep, is in a different physical state. We mentioned earlier that during sleep, body temperature remains normal and bodily functions proceed normally. In hibernation by contrast, there is a marked slowing down of body functions. In certain hibernating species the heartbeat can be as low as two beats per minute. Respiratory rates fall in direct proportion, breathing becoming very shallow. The temperature of the body can drop to close to zero, which explains the similarity of freezing and hibernation and the evolution of the term "cryogenic hibernation." Despite this considerable drop in body temperature the normal metabolic processes proceed at a sufficient rate to maintain the body of the animal at a temperature some 0.5° to 3° above that of the surrounding environment.

We must now consider how hibernation might enable men and women to reach the planets of "nearby" stars within their lifetime—or, putting it another way, how hibernation could stretch out their life-span by slowing down their life processes. Due to the greatly reduced metabolic rate, hibernating space voyagers would not require much feeding. Ideally they would not require any, but this seems unrealistic and presumably a slow drip feed would be essential. If the human body is cooled to 20°C below normal, the metabolic rate drops to less than 25 percent of normal. If genuine hibernation could somehow be induced, this would be reduced even further.

At present the process of hibernation is only vaguely understood. Prior to the onset of hibernation it is essential for hibernating creatures to lay in a considerable store of suitable food on which they gorge themselves prior to hibernation. Just why some animals have the capacity to hibernate while others, closely related, do not is a mystery. It is quite likely that the ability is controlled genetically.

Might some form of genetic engineering enable men and women desirous of crossing the interstellar wastes to hibernate? Genetic engineering, though still in its genesis, already is a highly emotive subject, and those scientists carrying out the first tentative experiments are frequently accused of playing God. This is not surprising, for clearly there are dangers and man might be taking the lid off a Pandora's box. However, when he succeeded in releasing the power of the atom he did just that, and it is a safe bet that nothing will inhibit him from doing it again, for man, by nature, is a questing creature. We should remember that while nuclear science has brought terrible dangers it also has brought considerable blessings in a number of fields, and the same could be true of genetic engineering.

Since the genesis of life on Earth, evolution has been based on the process of natural selection—that is, by naturally occurring mutations. With the advance of genetic engineering there could and probably will come a day when we can control our own evolution. We cannot predict what will happen then; we can only speculate. We could, perhaps, produce strains of humanity capable of surviving and developing in alien environments. If this were so then we might eventually be able to colonize worlds totally alien to us. But of greater interest to us is the possibility that we can produce human beings like ourselves who had been rendered natural hibernators. That would be one of the things that would open the way to the stars for us.

Hibernation does not halt the aging process but extends the lifetime because normal wear and tear have been greatly reduced. This might prove insufficient for interstellar travel, since the virtual negation of aging is required. A journey to Alpha Centauri at 0.02C (one-fiftieth the velocity of light) would take about 215 years. Stars several times as distant might involve journeys of 4 to 6 centuries. Lest the reader regard 0.02C as an inordinately low velocity, note that it represents roughly 3,700 miles per second, which is not exactly crawling. Thus with a form of hibernation that merely retarded the aging process the young men and women who left Earth to colonize the planet of another star would still have aged substantially on emerging from hibernation.

At present it is believed that hibernation could extend the human life-span by a factor of five. If we consider the average present life-span as 70 years, we would now have human beings with an effective life-span of 350 years. That might just be all right for a journey to Alpha Centauri, but for stars more remote it would not prove acceptable. A clear advantage of hibernation over some form of freezing would be the relative ease with which the space voyagers could be resuscitated.

Hibernation is a form of self-induced hypothermia from which no harm to the body results. It has distinct advantages over other postulated techniques of interstellar travel. With "generation travel" only distant descendants of the original voyagers reach the destination planet. Travel at near-light velocities (for example, 0.9C) may never prove feasible, and were it to become so we cannot be unmindful of the bizarre and disturbing effects of time dilation. At 0.02C these are insignificant.

So where then do we stand? At present there is no possibility of inducing hibernation in human beings but, in another century or two, genetic engineering may have solved this problem. Of course, the hibernation required would—unlike the hibernation we know among animals on earth—have to extend over two or three centuries. This represents a tremendous extrapolation of the process, yet it still represents the most plausible method we have for colonizing the planets of other stars.

Interstellar travel conducted in this manner creates images that are both poignant and macabre. Cleaving a silent path between Sun and stars go great star ships guided and controlled by instruments, while in separate sealed compartments, the star travelers sleep a deathlike sleep from which awakening may take place 100, 200, or 300 years hence. Already, back on Earth, the contemporaries and loved ones of these potential inheritors of a new world will have lived and died. To them life brought death, but to the travelers seeming death means a future life.

It might prove prudent on each star ship to have a few of the occupants awake and on watch at any one time—perhaps a year of wakefulness for every fifty in hibernation. This would be for the purpose of increased safety. Automatic control and navigation are all very well, but any scheme or device conceived by mortal man is capable of going wrong despite the existence of sophisticated fail-safe and backup systems. As Murphy's Law states, if a thing can go wrong, it will! Of course, those awake will be steadily consuming vital food and water, breathing precious oxygen, and aging during each year they are awake. Those, however, are simply the costs for an extra margin of safety, and it seems likely that we will be willing to pay them.

Appendix II: Toward a Planetary Home

IN CHAPTER 2 THE MATTER of detecting the presence of a planet that might have a capacity for terrestrial colonization was to some extent glossed over. It is time to rectify this deficiency. Even today, we can sometimes detect the existence of planetary systems around other stars, but determining with precision the nature of the atmosphere and the other physical parameters of these planets is not yet possible. Let us consider the means by which planetary systems are located.

It is possible at present to detect other planetary systems by either indirect or direct techniques. Indirect methods involve either astrometry or spectroscopy. In both those methods the existence of planets orbiting a star is inferred by the measurable effects a planet or planets have on that star. The direct method involves observation of electromagnetic radiation from the orbiting body or bodies themselves. Such radiation could take several forms:

- thermal radiation
- visible light from the parent star reflected by the planet or planets
- pulses of nonthermal radiation, such as those emitted by Jupiter

Astrometric and spectroscopic observations would appear to offer the most promise at the present time. Should a star possess a planetary companion (and in this instance we mean a large, massive planet), the star's orbital motion will revolve about the common center of gravity of both star and planet (usually referred to as the barycenter). It is theoretically possible to detect the presence of the large planetary companion by meticulous observation of the star's motion over a protracted period. The star's proper motion through space develops a "wobble." This technique enabled Peter van de Kamp to claim the existence of a large, dark, companion body orbiting Barnard's Star during the early summer of 1963. Some doubt has since been shed on this claim, but observational work is continuing. Even if van de Kamp's claim is fully substantiated (and chances are that it will be), the planet concerned would be a high-mass one and therefore totally unsuited for habitation by humans. Indeed, it would probably be unsuited for any form of living creature because of its extremely high gravity alone. There is a tendency to believe, however, that the presence of one (or more) large, high-mass planets also could mean the existence of a number of smaller ones. Whether these would be Earth-like or more akin to Mars or Venus is impossible to say. Barnard's Star is certainly not of the Sun's type, being considerably cooler than it, and a planet suitable for habitation by terrestrials would not only have to be of similar dimensions to Earth but also have to orbit Barnard's Star much more closely than Earth does the Sun. With this requirement fulfilled, the laws of celestial mechanics might well then force the planet to keep one hemisphere eternally turned to the star. This would hardly prove satisfactory. At this point it is logical to ask whether the Solar System is fairly typical of planetary systems or whether it tends toward the unique. We don't know, though the odds are it is not unique.

The spectroscopic indirect technique involves a search for those changes likely to occur in the spectrum of a star due to a planet or planets causing variation in the star's proper motion. Thus both astrometric and spectroscopic techniques depend on slight variations in the star's motion through space which, were the star unattended by companion bodies, would be a straight line. So far as spectrum shift is concerned, radiation emanating from a moving object (in this case a star) having a fluctuating path (a "wobble") should show a certain degree of Doppler shift.

In other words, to an observer on Earth the radiation would appear either redder (a shift to longer wavelengths) as the star receded, then bluer (a shift to shorter wavelengths) as it approached. The terms "receded" and "approached" in this context are, of course, very relative for, in relation to the distance separating the terrestrial observer from the star, they are minuscule. These alternating shifts would not occur if the star were following a rigid, straight-line course. It was by this principle that, in 1914, the late Edwin Hubble at Mount Wilson showed that the galaxies, hitherto regarded as objects within the Milky Way, were in fact separate entities lying well beyond it and receding at incredible velocities—that is, they displayed a Doppler shift to redder or longer wavelengths. The inescapable inference from this recession of the galaxies was that the universe was expanding. In like manner the observed spectrum of a star will change periodically from blue to red and back again as the star, due to its "wobbling" motion, alternately approaches and recedes from the observer. By recourse to highly accurate and painstaking spectroscopic observation, the degree of Doppler shift can be measured. This enables an estimate of the orbital speed of the star around the common center of mass of the star/planet system to be deduced. The effect can be produced only by a large, massive planet such as Jupiter (some astronomers question whether even Jupiter would prove large enough to show the effect). It is possible, therefore, even if extremely ironical to us, that astronomers on planets of other stars might regard the Sun as a lonely star, wholly devoid of planets.

Until very recently spectroscopic techniques could measure the speed of a star to within 1,000 meters per second, a figure that could hardly be described as excessive. Recently it has been found possible to improve this to a few hundred meters per second. These figures are not impressive, which is one of the reasons astrophysicists have not so far had much success using this method. If accuracy has already been increased from 1,000 meters a second to a few hundred, it is very likely they will be doing still better a century from now.

Astrometric observations are a better option at the present time. So high is the accuracy using this technique that, in theory at least, a Jovian-type planet (or at least its effects) could be detected orbiting some of the "nearby" stars. Unfortunately, we cannot categorically state that any have yet been found, though some are suspected. There could be two possible reasons why none has been definitely found: either these stars do not possess such large planets in their planetary systems, or the large planets have considerably smaller orbits and, being thus closer to their parent star, escape detection.

The first reason seems too facile. The second probably has the better pedigree. Current cosmological theories regarding the formation of planetary systems tend to view the existence of large gas-type planets like Jupiter, Saturn, Uranus, and Neptune as almost axiomatic. While the astrometric technique presently has a clear advantage over the spectroscopic, the astrometric technique still tends to be inadequate for a really comprehensive search of even the nearest stars. But these are early days yet.

Observational problems associated with the *direct* detection of other planetary systems are extremely severe. The greatest difficulty springs from the fact that stars are by their nature so much brighter than any of the planets likely to surround them. Planets shine only by virtue of the starlight reflected from them. They are extremely small in comparison to their parent stars, and no matter how high their albedos (reflecting power) they cannot possibly hope to compete with the great glaring luminaries they orbit. For example, in the Solar System we find that the Sun is about 2 billion times brighter than the giant planet Jupiter. An astronomer who happened to be observing our Sun from a planet of another star 33 light-years distant would, if he or she were lucky, perceive Jupiter only some 0.5 second of arc distant from the Sun. The space telescope designed by NASA for placement in orbit by the Space Shuttle will possess this degree of resolution. Unfortunately, it will not be able to cope with the contrast in brightness between star and planet to which we have already referred. However, all is not lost. The intrinsic thermal radiation emanating from Jupiter is strongest in the infrared portion of the spectrum. Here too, of course, the Sun still leads in the radiation stakes but only by a factor of about 10,000, which compares very favorably with the previously quoted figure of 2 billion.

Theoretically direct detection is superior to indirect detection inasmuch as direct detection takes a smaller fraction of the orbital period to obtain definitive results. Thus far we have not referred to determining the physical characteristics of any extrasolar planet discovered, and with regard to the concerns of this book, that is the area of greatest interest to us. Obviously this is even more difficult than detecting extrasolar planets, but, theoretically at least, we should be capable of making certain discoveries. In fact, indirect detection should enable the mass of a planet to be determined as well as its orbital period (its year) and its orbital inclination. Direct detection for its part could, if the infrared portion of the spectrum were used, provide some idea of a planet's temperature and atmospheric composition. It must be emphasized that this refers to large planets, the type most likely to be detected, the type that could not permit human visitation, far less habitation. This is not very promising, but this branch of astronomy is still in its infancy. What has been achieved so far has been in two decades or three. We must wonder what another century or two will render possible. It is a tantalizing thought.

Recent spectroscopic developments indicate that in the not too distant future it should be possible to determine the radial velocity of many stars with an accuracy of 10 meters per second, and, in time, this may be further improved to about 1 meter per second. Astrometric developments are also expected to continue. By the use of photoelectric detectors accuracies of a few million seconds should become realizable using existing telescopes. Some astronomers believe that within a few years ground-based astrometric observation might permit relative positional measurements accurate to 0.1 milliarc second.

If such accuracies can be anticipated from ground-based instruments, what can be expected from spaceborne telescopes? At present it is considered these would be marginally better than current ground-based observations but inferior to ground-based observations by the time space telescopes actually become operational. To rectify this anomalous position specially designed spaceborne astrometric systems will be required. If these can be provided—and there seems no reason why, in time, they should not—then accuracies up to 1 micro-arc per second could be attainable.

If adequate funds and a reasonable amount of time and thought are devoted to the search for extrasolar planetary systems, it should be possible in time to pinpoint those nearer stars having planetary families. To determine something of the physical nature and surface conditions will be a greater problem. It will be difficult even in the case of large Jovian-type planets. To determine the surface conditions of Earth-type planets is hardly likely to prove possible, since it will probably be impossible even to detect these planets. Colonists, it seems, must simply take a chance and go. Why not send instrumented probes? the reader may be tempted to ask. The reasons are plain: time and distance. But that question will be the subject of our third and concluding appendix.

Appendix III: Probing for Planets

FOR A LONG TIME NOW unmanned, instrumented probes have been spoken of as a method to tell us where other planetary systems exist and what the surface conditions on their planets are likely to be. It is a strangely unrealistic approach. Probes could be sent. But how long would they take to reach their destinations and signal back their findings to would-be colonists? A few specific examples will illustrate the position far better than a profusion of words.

1. Destination star: Alpha Centauri. Distance from Sun: 4.3 light-years. Speed of unmanned probe (C = light velocity): 0.02C.

$$\text{Time in transit} = \frac{4.3 \text{ light-years}}{0.02C}$$

(1 light-year is approximately 6 million million miles—that is, 6×10^{12} miles; C = 186,000 miles per second or 300,000 kilometers per second)

$$\text{Time in transit} = \frac{4.3 \times 6 \times 10^{12} \text{ miles}}{0.02 \times 186 \times 10^{3} \text{ miles sec}^{-1}} = \frac{25.8 \times 10^{9} \text{ sec}}{3.72}$$

$$= \frac{6.93 \times 10^{9}}{60 \times 60 \times 24 \times 365} \text{ years} = 200 \text{ years}$$

2. Destination star: Tau Ceti. Distance from Sun: 10.2 light-years.

$$\text{Time in transit} = \frac{10.2 \times 6 \times 10^{12} \text{ miles}}{0.02 \times 186 \times 10^{3} \text{ miles sec}^{-1}} = \frac{61.2 \times 10^{9} \text{ sec}}{3.72}$$

$$= 16.45 \times 10^{9} \text{ sec}$$

$$\frac{16.45 \times 10^{9}}{60 \times 60 \times 24 \times 365} \text{ years} = \frac{16.45 \times 10^{2}}{3.1536} \text{ years}$$

$$= \frac{1645}{3.15} \text{ years} = 522 \text{ years}$$

3. Destination star: Epsilon Indi. Distance from Sun: 11.6 light-years.

$$\text{Time in transit} = \frac{11.6 \times 6 \times 10^{12} \text{ miles}}{0.02 \times 186 \times 10^{3} \text{ miles sec}^{-1}} = \frac{69.6 \times 10^{9}}{3.72}$$

$$= 18.71 \times 10^{9} \text{ sec}$$

$$= \frac{18.71 \times 10^{9}}{3.15 \times 10^{7}} \text{ years} = \frac{18.71 \times 10^{2}}{3.15} \text{ years}$$

$$= 594 \text{ years}$$

These three stars are among the closest to the Solar System, and their stellar classifications are either the same as that of the Sun or closely akin to it. Yet at a velocity of 0.02C unmanned instrumented probes would take from 200 to 600 years to reach them. These are lengthy periods for colonists to have to wait! It could, of course, be argued that 0.02C is not a particularly high velocity. However, 0.02C is equivalent to 3.720 miles per second or 13.39 million miles per hour! Even if probe velocities could be increased threefold, the probes still would take 60 to 200 years to reach their destinations. Surely even the most optimistic must agree that time scales of this order are not practical. Only if some of the extradimensional possibilities outlined in *Interstellar Travel: Past, Present, and Future* could be realized, giving very brief transit times, would such probes have practical value. But if these possibilities do come to pass, it would be just as sensible to dispatch manned reconnaissance ships. Even if extradimensional travel is feasible, it will be long before humans perfect it.

The possibility of travel at near-light velocities is another question but one in which the difficulties speak loudly for themselves. Indeed, they virtually shout. Travel at the velocity of light is not possible, and as we will seek to show mathematically, travel beyond the velocity of light is an apparent absurdity. To reach near-light velocities would require power sources beyond our present imagining, and even assuming these could be achieved, the more remote stars would still lie well outside our range—or more correctly our lifetimes. Certainly near-light velocities (or near-optical velocities, as they are more usually referred to) would have a modicum of relevance in the case of the nearer stars already mentioned: 4.3 years to Alpha Centauri, 10.2 years to Tau Ceti, and 11.6 years to Epsilon Indi. But even 4½ to 5 years traveling in space is a reasonably lengthy period by human standards, almost 12 years even more so. There would, however, be a bonus, and this seems a good place to illustrate it by simple mathematics. All this, of course, refers to reconnaissance vessels (human probes).

As a consequence of relativity it can be shown that an increase in velocity has the effect of slowing time, an effect generally described as time dilation. Thus a chronometer aboard a star ship would run more slowly than one on Earth. The effect would coincide with the retardation of all biological, physical, and chemical processes in the bodies of the men and women aboard the star ship, not that this would be apparent to them. So far as they are concerned, all would seem perfectly normal.

The time intervals between two events as measured by these two clocks are directly related, and this relationship can be expressed in the form of an equation as follows:

$$\frac{T_{ss}}{T_t} = \sqrt{1 - \frac{V^2}{C^2}}$$

T_{ss} represents the time interval recorded on the star ship's clock, T_t that recorded by the terrestrial clock. V is the velocity of the star ship relative to Earth. C is the velocity of light.

Consequently, if the velocity of the star ship V is low, then the right-hand side of the equation tends in value toward unity, in which case T_{ss} is equal to T_t; in other words, the time interval recorded by the two clocks is the same, and that is surely what we would expect. However, if the velocity of the star ship is high though still well below that of light, a disparity begins to manifest itself, for then $\frac{V^2}{C^2}$ tends toward a small but nevertheless tangible value. The right-hand side of the equation has then a value slightly less than unity, in which case T_{ss} becomes equal to a fraction of T_t, so that a time interval aboard the star ship is slightly less than the corresponding time interval on Earth. Perhaps by now it will be evident where all this is leading. Let us now consider the extreme case where the velocity of the star ship is very close to that of light. $\frac{V^2}{C^2}$ then becomes almost equal to unity,

so the right-hand side of the equation, $\sqrt{(1 - \frac{V^2}{C^2})}$, becomes

infinitesimally small. Consequently T_{ss} is equal to a very small fraction of T_t—that is, a time interval on the star ship is now very much less than the corresponding time interval on Earth.

We said earlier that the attainment of light velocity is impossible, though it is theoretically possible to reach a velocity of 0.99C. Using the above equation for light velocity we find that in this instance $\frac{V^2}{C^2}$ equals unity and consequently the entire right-hand side of the equation becomes 0. If we accept this, then $\frac{T_{ss}}{T_t}$ equals 0, in which case T_{ss} also must equal 0. For a velocity exceeding that of light, the position becomes even more ludicrous. If we assume, for the sake of argument, that V equals 1.1C, then $\frac{V^2}{C^2}$ is greater than unity, in which case the right-hand side of the equation becomes equal to a minus quantity. Thus T_{SS} is equal to $-T_t$, so that a time interval on the star ship is equal to a negative interval of terrestrial time, which, on the face of it, is nonsensical. However, it might be as well to reserve judgment in view of the comments we will make on tachyon travel.

Time dilation could have practical value to space voyagers only if velocities close to that of light could be achieved. Once again, a few mathematical examples may assist.

1. Star: Proxima Centauri. Distance from Solar System: 4.2 light-years. Speed of star ship: 0.66C or 200,000 kilometers per second.

$$\frac{T_{SS}}{T_t} = \sqrt{1 - \frac{V^2}{C^2}}$$

By substitution, $$\frac{T_{SS}}{T_t} = \sqrt{1 - \frac{(0.66C)^2}{C^2}} = \sqrt{1 - \frac{0.436C^2}{C^2}}$$

$$= \sqrt{1 - 0.436} = \sqrt{0.564} = 0.751$$

Therefore, $T_{SS} = 0.751\ T_t$

Thus if the journey could be made at light velocity C it would occupy 4.2 years but according to the equation *no* time at all. $T_{SS} = 0$. $T_t = 0$. This is absurd and occurs because the velocity of light C is unattainable. However, if it is made at 66 percent of light velocity the time taken should be one-third longer—that is, 5.6 years. Therefore,

$T_{SS} = 0.75 \times 5.6 = 4.2$ years

Consequently, those remaining on Earth would have aged by 5.6 years, those on the star ship by only 4.2 years—an advantage of 1.4 years. Not a great deal, perhaps, but nevertheless a tangible difference.

2. Star: Procyon. Distance from Solar System: 10.4 light-years. Speed of star ship: 0.99C.

$$\frac{T_{ss}}{T_t} = \sqrt{1 - \frac{V^2}{C^2}} = \sqrt{1 - \frac{(0.99C)^2}{C^2}} = \sqrt{1 - 0.98} = \sqrt{0.02}$$

$$= 0.14$$

$$T_t = 10.4 \text{ years; therefore, } T_{ss} = 10.4 \times 0.14 = 1.5 \text{ years}$$

The time of the journey is 1.5 years to the star ship's occupants. The aging difference therefore is 10.4 - 1.5 years, or approximately 9 years.

It can be seen that reconnaissance missions to investigate the planetary systems of some of the nearer stars apparently are not impossible. This would assuredly take much of the chance out of colonizing missions, *but* very high submultiples of light velocity would need to be attained. For the 1½-year voyage to the star Procyon in the constellation Canis Minor this would be 0.99C, almost the speed of light itself. Could such a fantastic velocity ever be attained? Let us look at a few startling facts. It has been estimated that for a very small space vessel weighing a mere one ton, capable of traveling at 0.8C, the energy requirements would be about 215 billion kilowatt-hours, or the total output of all the generating stations on Earth over a period of several months. It is difficult to say what degree of credence should be given to estimates of this nature, but chances are that this one is not very wide of the mark. We must also face two additional unpalatable facts. At velocities of this order very serious and complex navigational problems would almost certainly arise and the smallest specks of cosmic dust lying in the path of the vessel could have a lethal impact. These difficulties might eventually be obviated but one suspects a great deal of time would be required. Human beings desperate to find alternative worlds to Earth might not wait that long.

Is the speed of light really a bounding velocity? Certainly the mathematics involved clearly indicate it is. Suppose, for example, we have a star ship moving with velocity V in a fixed system of reference. If that velocity is increased by an increment U, then its overall velocity should become, as one would expect, V + U. By the relativistic laws of kinematics, however, the result of such a velocity composition W is not W = V + U but:

$$W = \frac{V + U}{1 + \frac{UV}{C^2}}$$

If V is equal to 0.9C and if U is equal to 0.1C, it would appear, if one uses conventional logic, that the final velocity should be C, that of light. However, this merely involves the expression W = V + U, which is invalid in the circumstances. Using the proper equation a strange, almost bizarre paradox begins to manifest itself. Thus:

$$W = \frac{0.9C + 0.1C}{1 + \frac{0.1C \times 0.9C}{C^2}} = \frac{C}{1 + \frac{0.09C^2}{C^2}} = \frac{C}{1.09}$$

Therefore, W = 0.918C

Similarly, we find that if we make the increment U equal to 0.2C, the resultant velocity is not 1.1C but 0.923C. In like manner, if we add nine increments of 0.1C, the effective star ship velocity is not doubled but is only 0.994C. It might appear that if we were to add ten such increments the final value must surely be C. Not so.

$$W = \frac{0.9C + 1.0C}{1 + \frac{0.9C \times 1.0C}{C^2}} = \frac{1}{1 + \frac{0.9C^2}{C^2}} = \frac{1}{1 + .09} \text{ or } \frac{C}{1.9}$$

Therefore, W = 0.526C

This would not be possible, since ten increments of 0.1C are equal to C, the velocity of light. And in any case (mathematically, at least), resultant velocity is down to just over half that of light. Irrespective of how many velocity increments are added, a star ship could never attain the absolute velocity indicated by:

$$W = C$$

though theoretically it would be possible to come fairly close to it.

Though this is all very convincing mathematically, a note of caution must be struck because of the postulated but still unproven existence of tachyons mentioned earlier. A tachyon is a theoretical nuclear particle believed to be moving faster than light. Tachyons might be more fully defined as particles moving with infinite velocities that, when approaching the speed of light, are slowing down. It is thought that tachyons might have velocities as high as a billion times the speed of light. When in slowing down they reach the velocity of light, they cease to exist.

At present tachyons are abstractions, though attempts to confirm their existence are proceeding. The difficulties are tremendous. If tachyons exist they are traveling beyond the speed of light, and therefore we cannot see them. If somehow we slow them down sufficiently they cease to exist, in which case we still cannot see them. The only really feasible approach to the problem is to try to detect them by virtue of their effect on some entity we *can* observe, but this approach is complicated by the fact that it is difficult to attribute any effect specifically to tachyons. The matter is discussed at some length in *Interstellar Travel: Past, Present, and Future,* to which the interested reader is referred.

The question of travel at hyperlight velocities by harnessing tachyons in some way can be assessed as follows:

1. The existence of tachyons must first be confirmed.
2. Means of harnessing them must be devised.
3. Since tachyons cease to exist at the velocity of light, any space vessel must first attain the velocity of light plus a small increment. But if the normal attainment of the velocity of light is impossible, there is a "velocity gap" that must somehow be overcome before any form of "tachyon drive" could take over.

4. Even if hyperlight velocities could be attained, we still have to consider relativistic effects such as time dilation and also navigational problems and the effect of even the smallest particles of solid matter in the path of the star ship.

It should be stressed that Einstein merely says that travel at the velocity of light is impossible, not that speeds in excess of it are impossible. Hence tachyons, if they do exist, can travel at velocities beyond that of light but not at it.

Unfortunately (or fortunately, depending on the point of view), tachyons, if they exist and can travel at velocities in excess of that of light, make nonsense of equations proving that hyperlight velocities are impossible.*

---

*In the mid-1970s, Edward Meier, a Swiss citizen, claimed to have had contact with humanoidlike extraterrestrial beings from a planet in the region of the Pleiades star cluster. These alleged beings explained their ability to traverse vast interstellar distances (the Pleiades are about 600 light-years away) by virtue of a "tachyon drive." This is the kind of UFO story that in normal circumstances probably would best be ignored. However, the claim was extensively investigated by Wendelle Stevens, a retired colonel of the U.S. Air Force. To further his inquiries he enlisted the services of an investigative unit in Arizona whose main business is industrial counterespionage. Their intention was to "crack" Meier by revealing the flaws in his story. However, after four years of close investigation that included very thorough computer analysis of the photographs taken, tests of metal samples and of the taped sound of the spacecraft, the investigative unit was unable to detect any flaws that would suggest a hoax. The entire episode became the subject of an illustrated book, *UFO—Contact from the Pleiades* (Phoenix, Ariz.: Genesis III Publishing) and later of a documentary film, *Contact* (Los Angeles: Savadove-Young Films), which has just been released in the United States. I was asked to take a minor part in this film to comment on the subject of tachyon hyperlight travel. The investigation is continuing and in view of the facts, the results might be very far-reaching. Only time will tell. Perhaps the universe is not merely strange. It may be a whole lot stranger than we can possibly imagine!

Index